HARCOURT

Math

Teacher's Resource Book

Grade 4

Orlando • Boston • Dallas • Chicago • San Diego
www.harcourtschool.com

Printed in the United States of America

ISBN 0-15-320949-6

1 2 3 4 5 6 7 8 9 10 018 2004 2003 2002 2001

CONTENTS

PROBLEM SOLVING

NUMBER AND OPERATIONS

TIME, MONEY, MEASUREMENT

GEOMETRY

DATA, PROBABILITY, AND GRAPHING

TEACHER'S EDITION PRACTICE GAMES

DAILY FACTS PRACTICE

FACT CARDS

VOCABULARY CARDS

TEACHER'S RESOURCE BOOK

This section includes various types of resources for lessons in *Harcourt Math.*

Resources are provided for the following categories:

- **Problem Solving**
- **Number and Operations**
- **Money, Measurement**
- **Data, Probability, and Graphing**
- **Geometry**
- **Teacher's Edition Practice Games**
- **Daily Facts Practice**
- **Fact Cards**

Name ______________________________

Problem Solving

Understand

1. Retell the problem in your own words. ______________________________

2. List the information given. ______________________________

3. Restate the question as a fill-in-the-blank sentence. ______________________________

Plan

4. List one or more problem-solving strategies that you can use. ______________________________

5. Predict what your answer will be. ______________________________

Solve

6. Show how you solved the problem. ______________________________

7. Write your answer in a complete sentence. ______________________________

Check

8. Tell how you know your answer is reasonable. ______________________________

9. Describe another way you could have solved the problem. ______________________________

Name ______________________

Problem Solving Think Along

Understand

1. What is the problem about?
2. What information is given in the problem?
3. What is the question?

Plan

4. What problem-solving strategies might I try to help me solve the problem?
5. About what do I think my answer will be?

Solve

6. How can I solve the problem?
7. How can I state my answer in a complete sentence?

Check

8. How do I know whether my answer is reasonable?
9. How else might I have solved this problem?

4	0
5	1
6	2
7	3

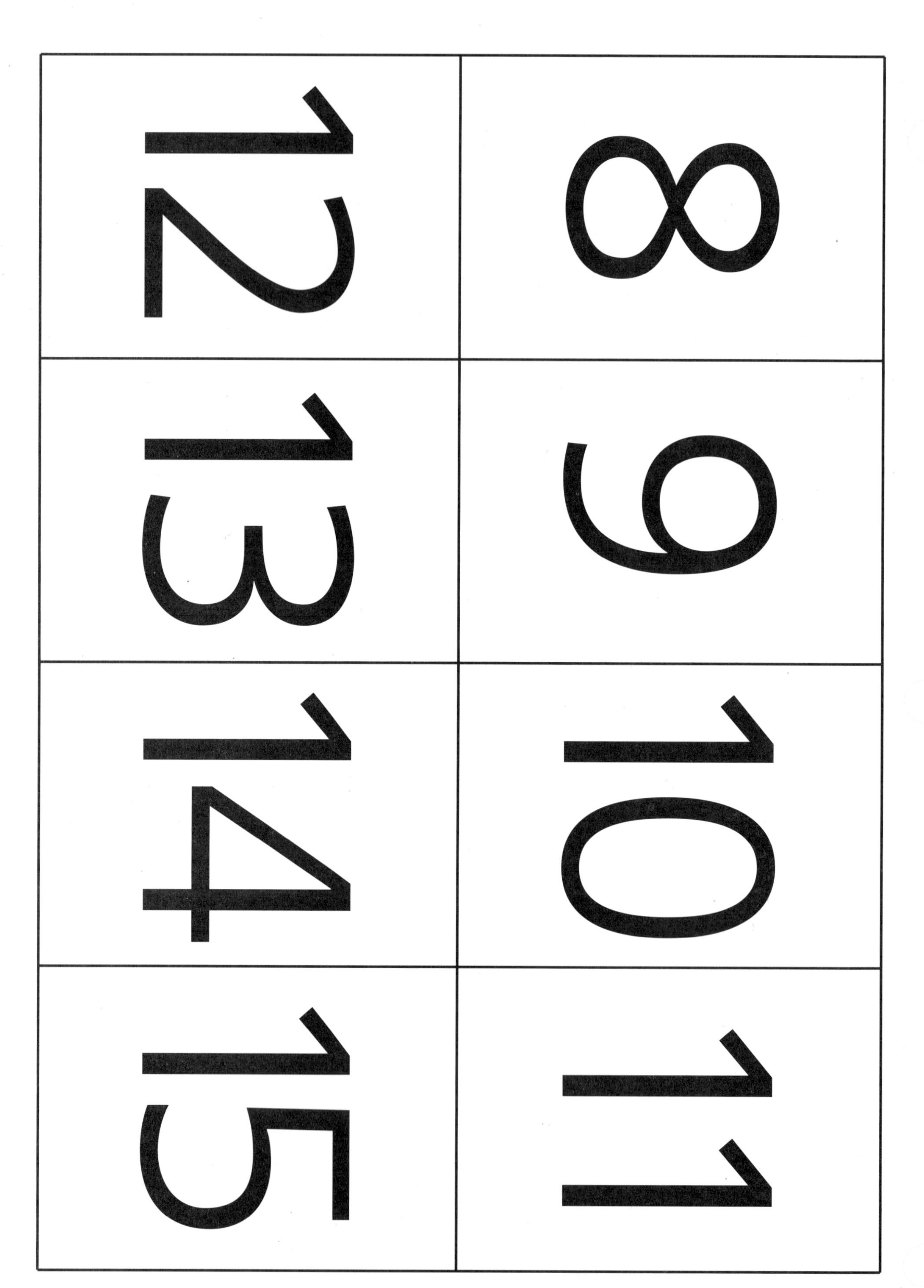
8
9
10
11
12
13
14
15

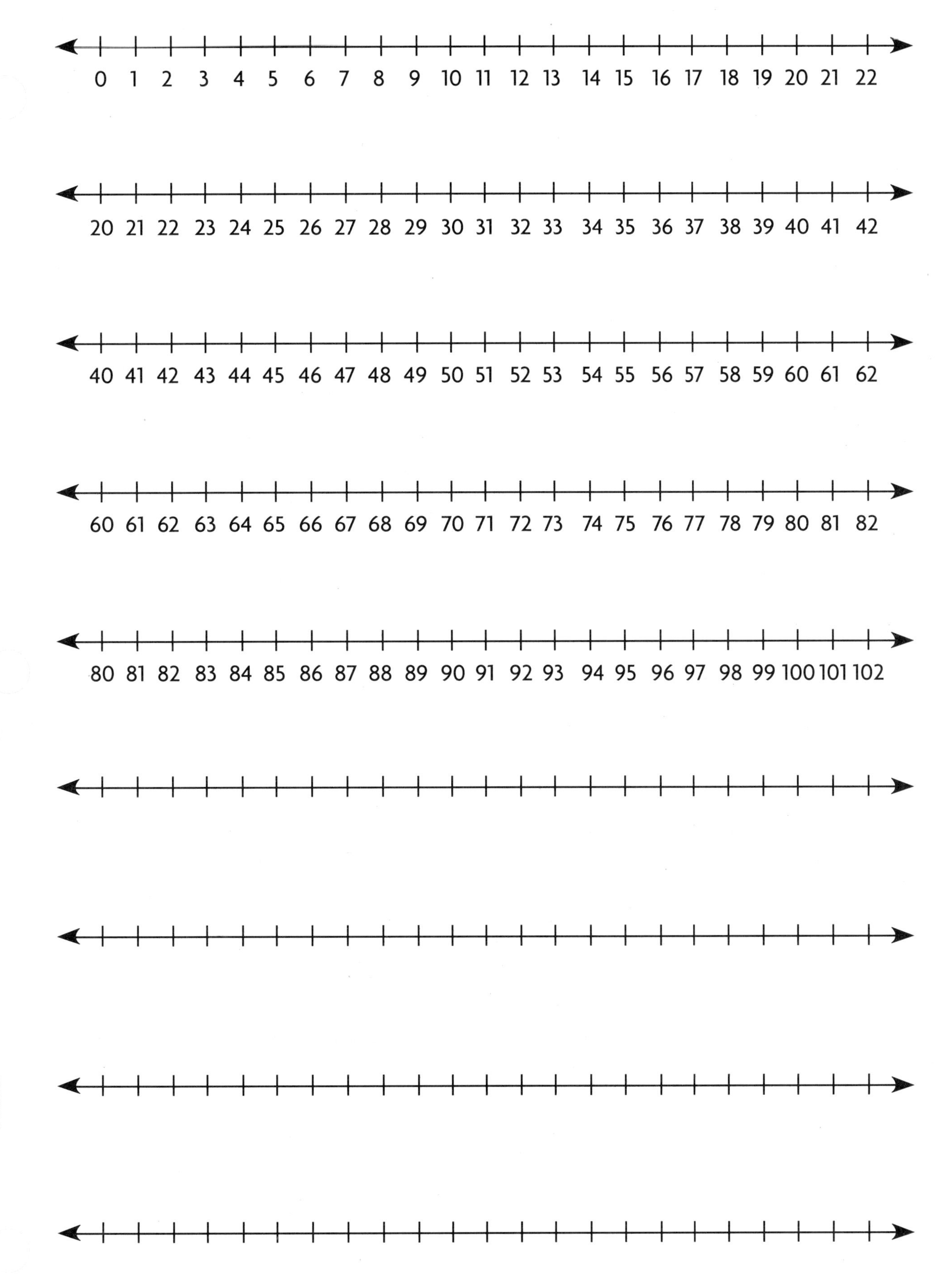
0 1 2 3 4 5 6 7 8 9 10 11 12 13 14 15 16 17 18 19 20 21 22
20 21 22 23 24 25 26 27 28 29 30 31 32 33 34 35 36 37 38 39 40 41 42
40 41 42 43 44 45 46 47 48 49 50 51 52 53 54 55 56 57 58 59 60 61 62
60 61 62 63 64 65 66 67 68 69 70 71 72 73 74 75 76 77 78 79 80 81 82
80 81 82 83 84 85 86 87 88 89 90 91 92 93 94 95 96 97 98 99 100 101 102

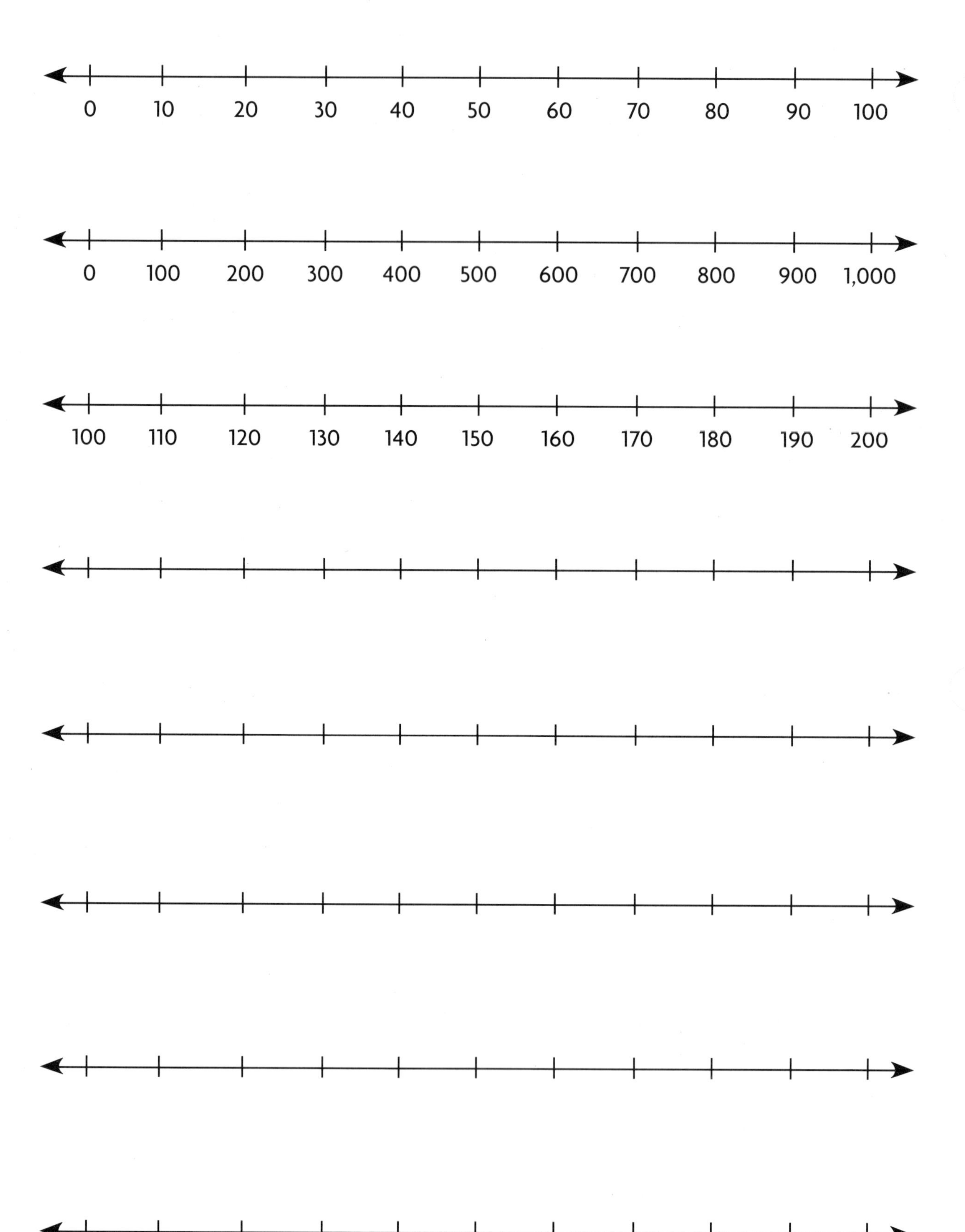
0
10
20
30
40
50
60
70
80
90
100
0
100
200
300
400
500
600
700
800
900
1,000
100
110
120
130
140
150
160
170
180
190
200

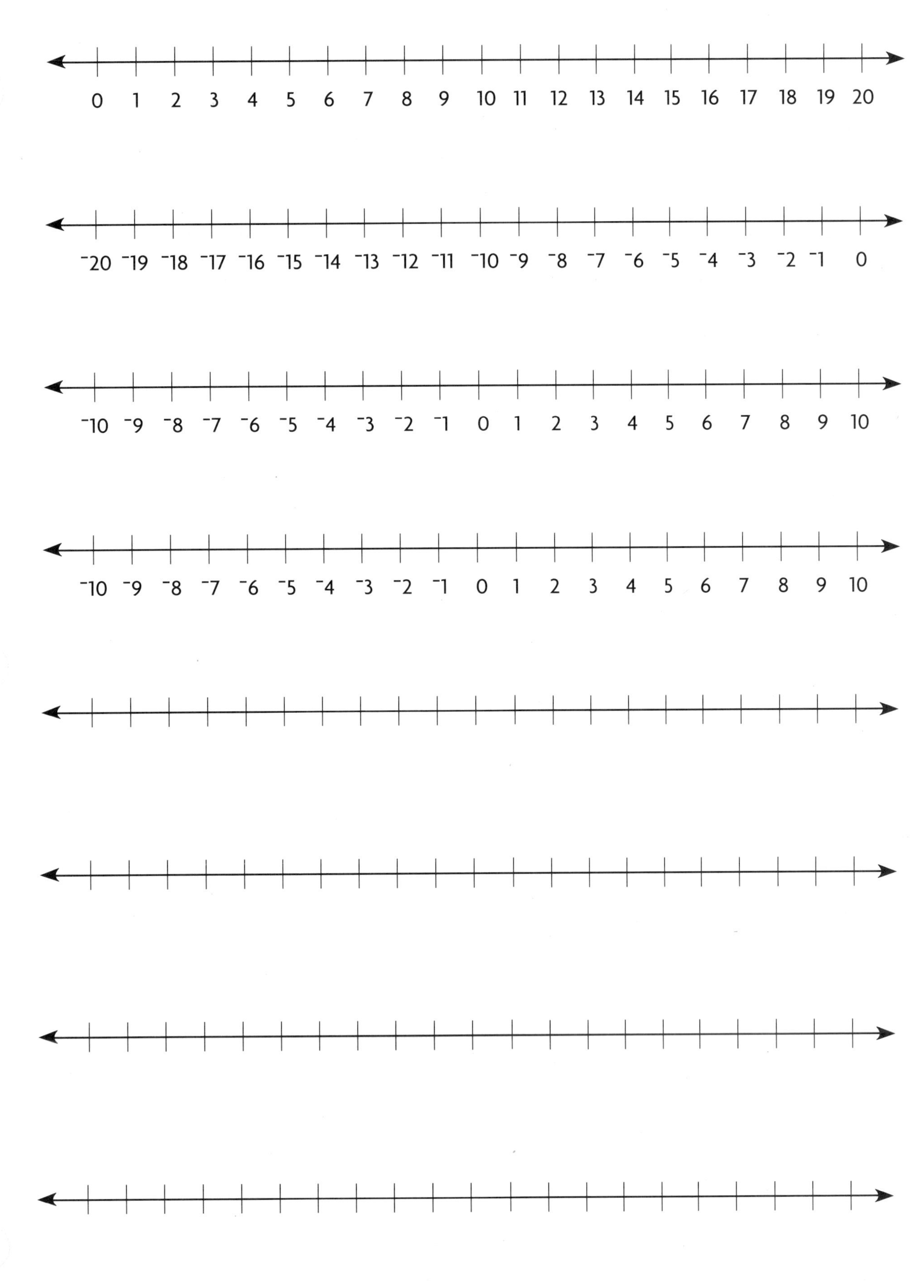
0 1 2 3 4 5 6 7 8 9 10 11 12 13 14 15 16 17 18 19 20
-20 -19 -18 -17 -16 -15 -14 -13 -12 -11 -10 -9 -8 -7 -6 -5 -4 -3 -2 -1 0
-10 -9 -8 -7 -6 -5 -4 -3 -2 -1 0 1 2 3 4 5 6 7 8 9 10
-10 -9 -8 -7 -6 -5 -4 -3 -2 -1 0 1 2 3 4 5 6 7 8 9 10

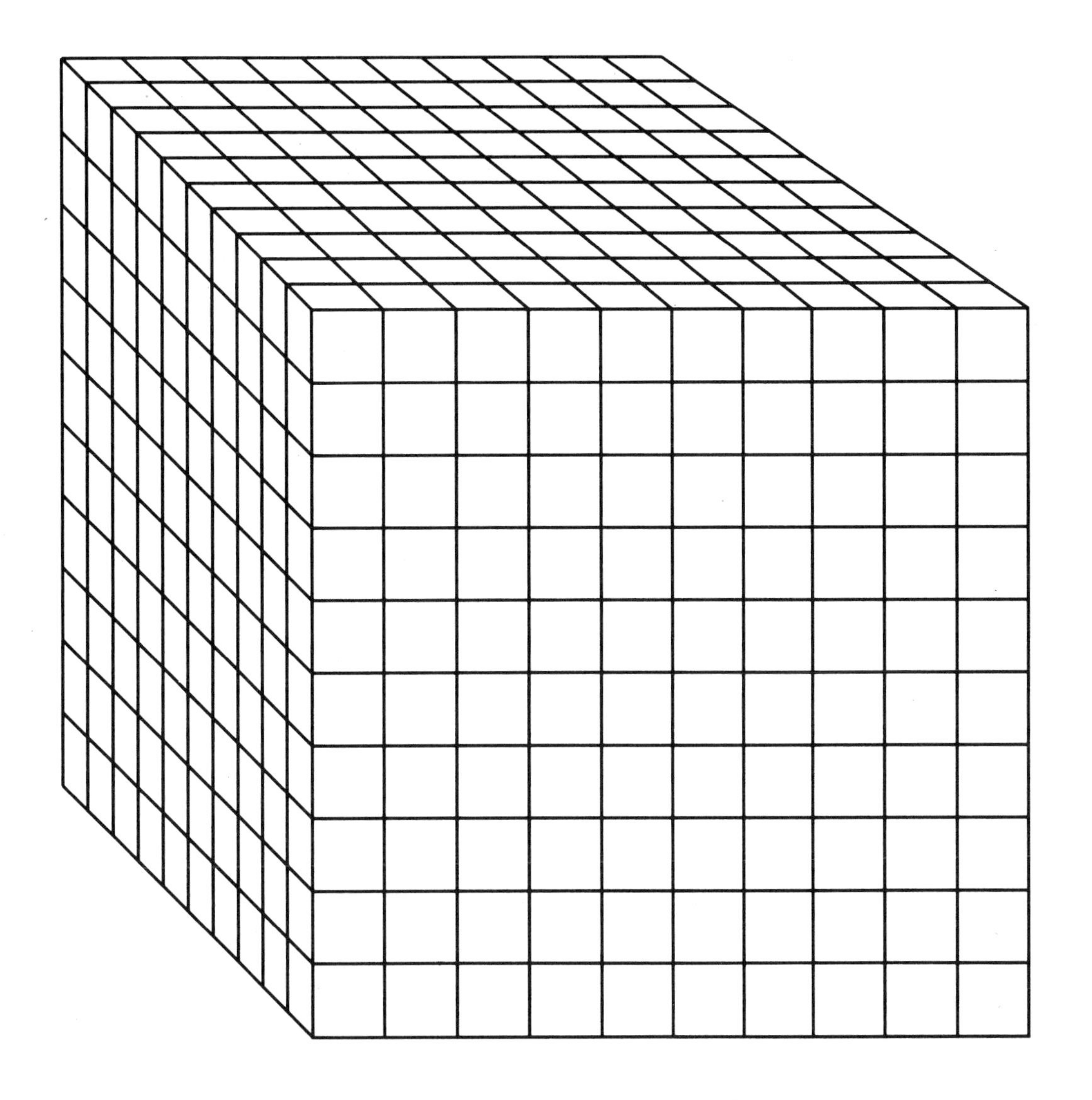

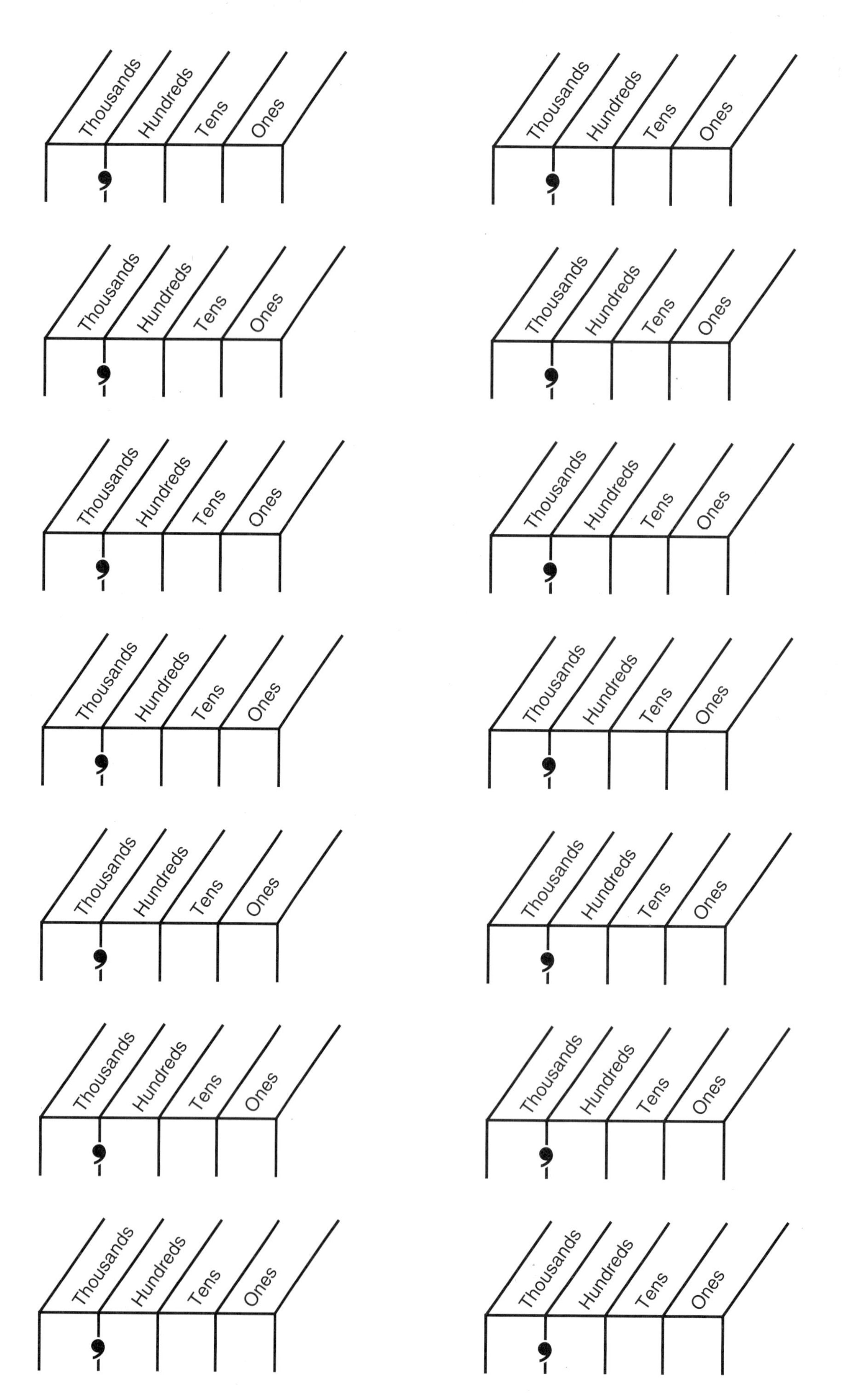
Thousands
Hundreds
Tens
Ones

Hundred Thousands	Ten Thousands	Thousands	Hundreds	Tens	Ones
		,			

Hundred Thousands	Ten Thousands	Thousands	Hundreds	Tens	Ones
		,			

Hundred Thousands	Ten Thousands	Thousands	Hundreds	Tens	Ones
		,			

Hundred Thousands	Ten Thousands	Thousands	Hundreds	Tens	Ones
		,			

Hundred Thousands	Ten Thousands	Thousands	Hundreds	Tens	Ones
		,			

Hundred Thousands	Ten Thousands	Thousands	Hundreds	Tens	Ones
		,			

Hundred Thousands	Ten Thousands	Thousands	Hundreds	Tens	Ones
		,			

Hundred Thousands	Ten Thousands	Thousands	Hundreds	Tens	Ones
		,			

Hundred Thousands	Ten Thousands	Thousands	Hundreds	Tens	Ones
		,			

Hundred Thousands	Ten Thousands	Thousands	Hundreds	Tens	Ones
		,			

Hundred Thousands	Ten Thousands	Thousands	Hundreds	Tens	Ones
		,			

Hundred Thousands	Ten Thousands	Thousands	Hundreds	Tens	Ones
		,			

Hundred Thousands	Ten Thousands	Thousands	Hundreds	Tens	Ones
		,			

Hundred Thousands	Ten Thousands	Thousands	Hundreds	Tens	Ones
		,			

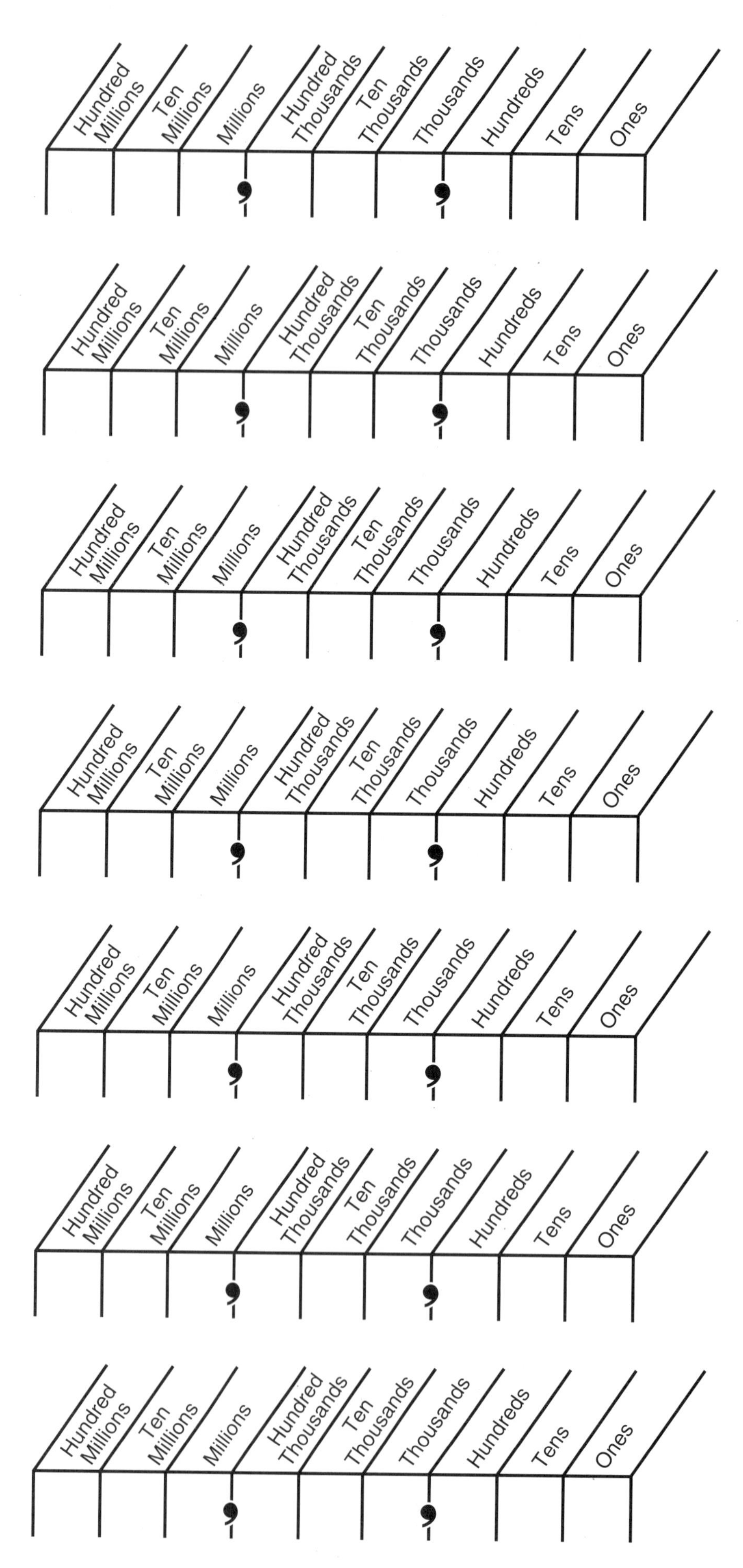
Hundred Millions
Ten Millions
Millions
Hundred Thousands
Ten Thousands
Thousands
Hundreds
Tens
Ones

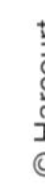

Thousands	Hundreds	Tens	Ones

Millions	Hundred Thousands	Ten Thousands	Thousands	Hundreds	Tens	Ones

	0	1	2	3	4	5	6	7	8	9	10	11	12
0													
1													
2													
3													
4													
5													
6													
7													
8													
9													
10													
11													
12													

×	1	2	3	4	5	6	7	8	9	10	11	12
1	1	2	3	4	5	6	7	8	9	10	11	12
2	2	4	6	8	10	12	14	16	18	20	22	24
3	3	6	9	12	15	18	21	24	27	30	33	36
4	4	8	12	16	20	24	28	32	36	40	44	48
5	5	10	15	20	25	30	35	40	45	50	55	60
6	6	12	18	24	30	36	42	48	54	50	66	72
7	7	14	21	28	35	42	49	56	63	70	77	84
8	8	16	24	43	40	48	56	64	72	80	88	96
9	9	18	27	36	45	54	63	72	81	90	99	108
10	10	20	30	40	50	60	70	80	90	100	110	120
11	11	22	33	44	55	66	77	88	99	110	121	132
12	12	24	36	48	60	72	84	96	108	120	132	144

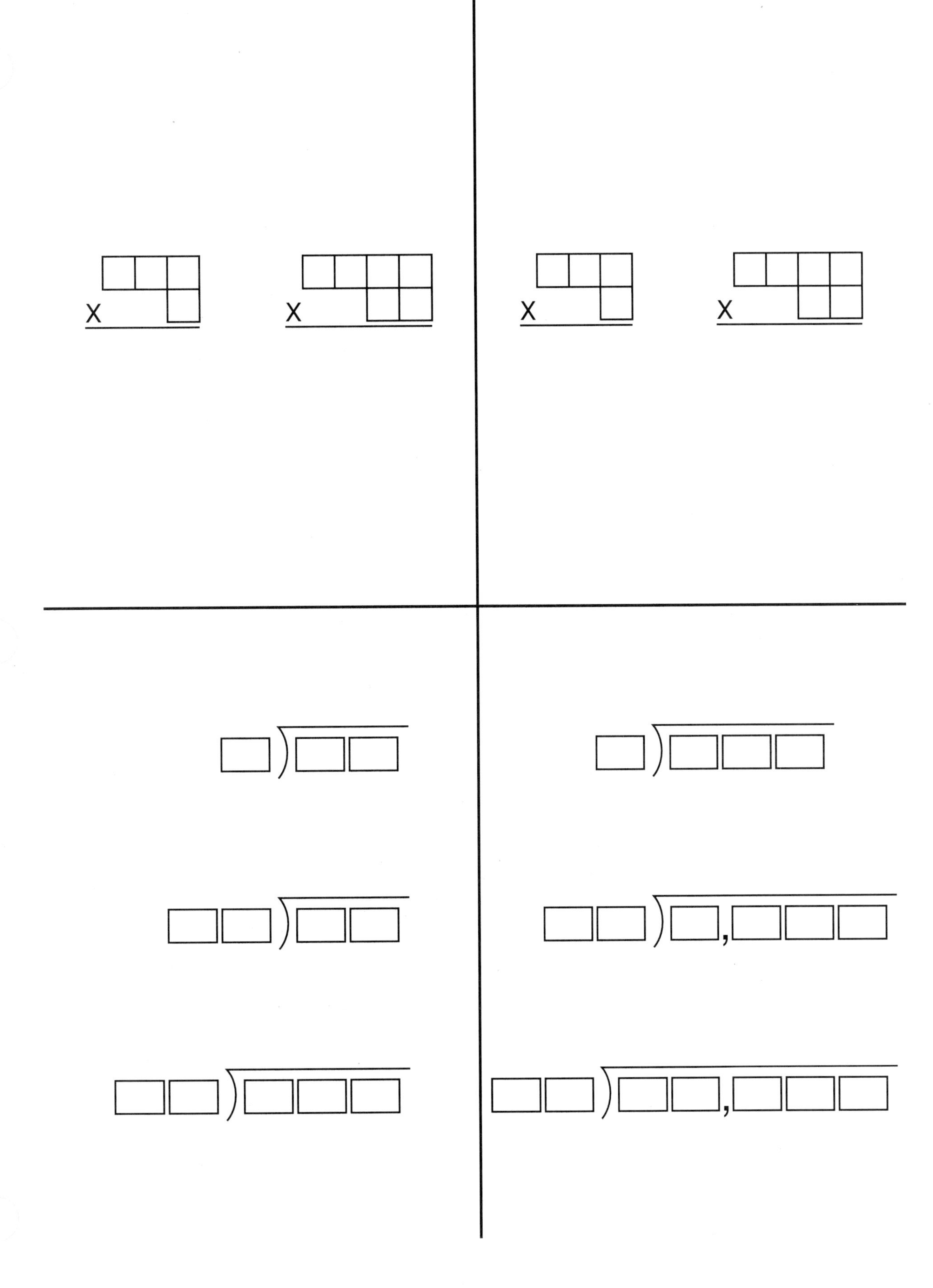
X
X
X
X

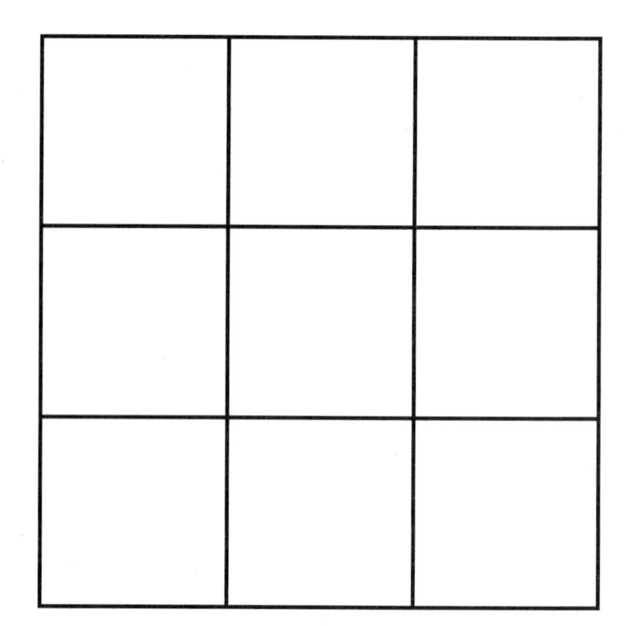

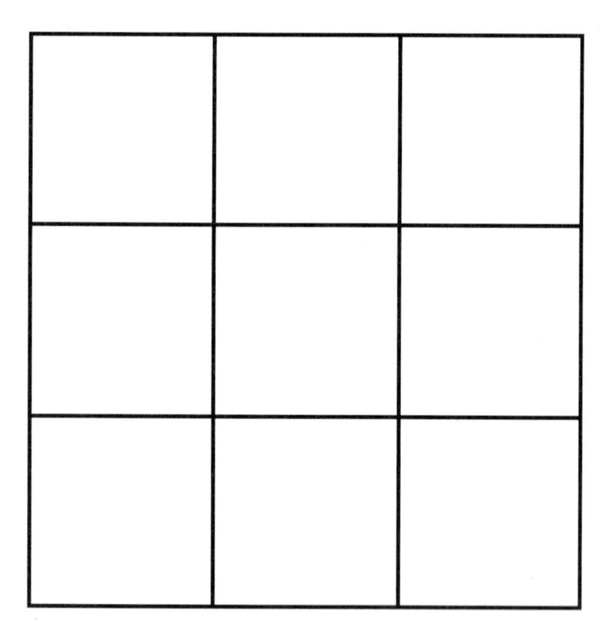

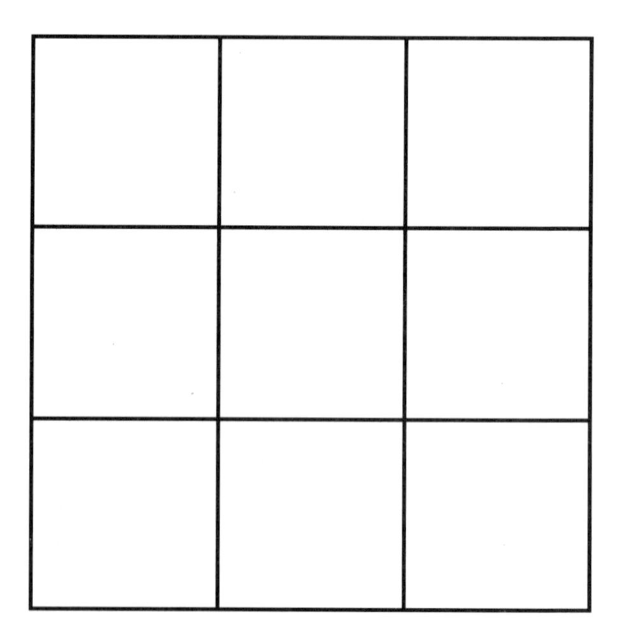

$\frac{1}{12}$	$\frac{1}{12}$	$\frac{1}{12}$	$\frac{1}{12}$	$\frac{1}{12}$	$\frac{1}{12}$	$\frac{1}{12}$	$\frac{1}{12}$	$\frac{1}{12}$	$\frac{1}{12}$	$\frac{1}{12}$	$\frac{1}{12}$

$\frac{1}{11}$	$\frac{1}{11}$	$\frac{1}{11}$	$\frac{1}{11}$	$\frac{1}{11}$	$\frac{1}{11}$	$\frac{1}{11}$	$\frac{1}{11}$	$\frac{1}{11}$	$\frac{1}{11}$	$\frac{1}{11}$

$\frac{1}{10}$	$\frac{1}{10}$	$\frac{1}{10}$	$\frac{1}{10}$	$\frac{1}{10}$	$\frac{1}{10}$	$\frac{1}{10}$	$\frac{1}{10}$	$\frac{1}{10}$	$\frac{1}{10}$

$\frac{1}{9}$	$\frac{1}{9}$	$\frac{1}{9}$	$\frac{1}{9}$	$\frac{1}{9}$	$\frac{1}{9}$	$\frac{1}{9}$	$\frac{1}{9}$	$\frac{1}{9}$

$\frac{1}{8}$	$\frac{1}{8}$	$\frac{1}{8}$	$\frac{1}{8}$	$\frac{1}{8}$	$\frac{1}{8}$	$\frac{1}{8}$	$\frac{1}{8}$

$\frac{1}{7}$	$\frac{1}{7}$	$\frac{1}{7}$	$\frac{1}{7}$	$\frac{1}{7}$	$\frac{1}{7}$	$\frac{1}{7}$

$\frac{1}{6}$	$\frac{1}{6}$	$\frac{1}{6}$	$\frac{1}{6}$	$\frac{1}{6}$	$\frac{1}{6}$

$\frac{1}{5}$	$\frac{1}{5}$	$\frac{1}{5}$	$\frac{1}{5}$	$\frac{1}{5}$

$\frac{1}{4}$	$\frac{1}{4}$	$\frac{1}{4}$	$\frac{1}{4}$

$\frac{1}{3}$	$\frac{1}{3}$	$\frac{1}{3}$

$\frac{1}{2}$	$\frac{1}{2}$

1

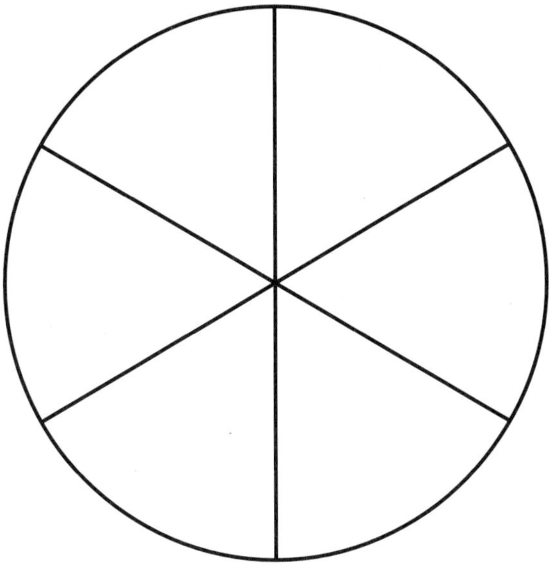

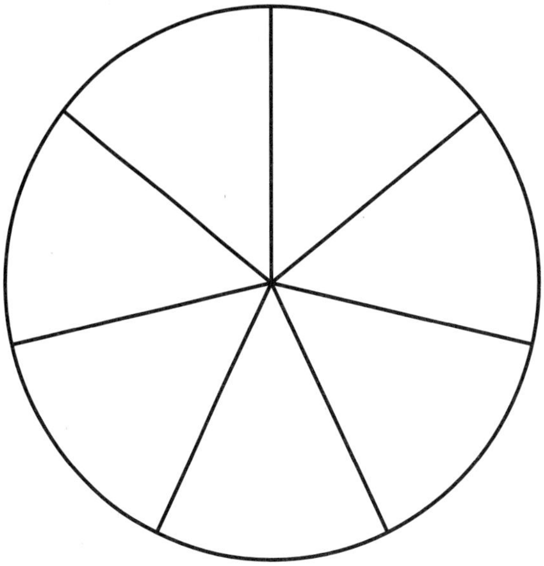

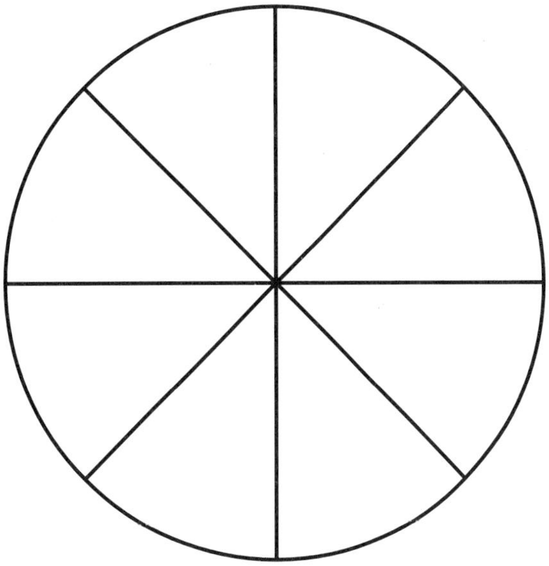

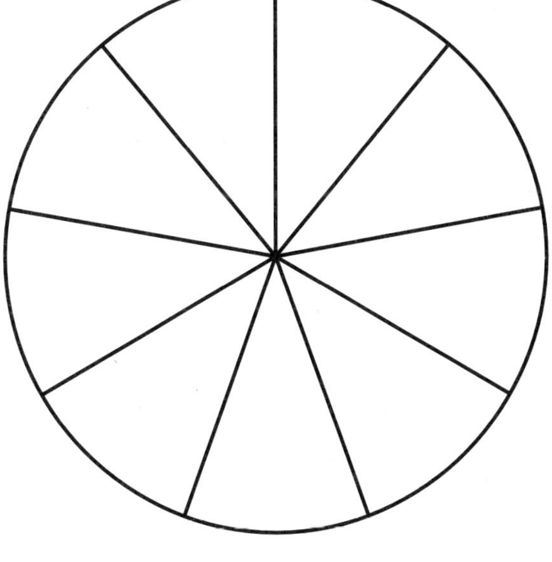

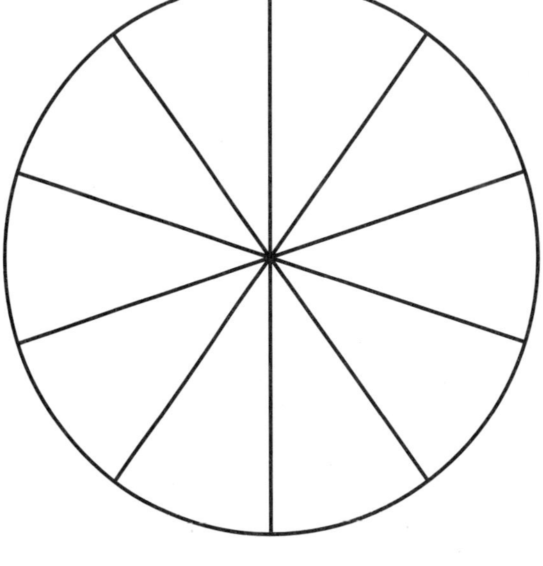

Fraction Circles

0 $\frac{1}{2}$ 1

0 $\frac{1}{3}$ $\frac{2}{3}$ 1

0 $\frac{1}{4}$ $\frac{2}{4}$ $\frac{3}{4}$ 1

0 $\frac{1}{5}$ $\frac{2}{5}$ $\frac{3}{5}$ $\frac{4}{5}$ 1

0 $\frac{1}{6}$ $\frac{2}{6}$ $\frac{3}{6}$ $\frac{4}{6}$ $\frac{5}{6}$ 1

0 $\frac{1}{8}$ $\frac{2}{8}$ $\frac{3}{8}$ $\frac{4}{8}$ $\frac{5}{8}$ $\frac{6}{8}$ $\frac{7}{8}$ 1

0 $\frac{1}{9}$ $\frac{2}{9}$ $\frac{3}{9}$ $\frac{4}{9}$ $\frac{5}{9}$ $\frac{6}{9}$ $\frac{7}{9}$ $\frac{8}{9}$ 1

0 $\frac{1}{10}$ $\frac{2}{10}$ $\frac{3}{10}$ $\frac{4}{10}$ $\frac{5}{10}$ $\frac{6}{10}$ $\frac{7}{10}$ $\frac{8}{10}$ $\frac{9}{10}$ 1

0 $\frac{1}{12}$ $\frac{2}{12}$ $\frac{3}{12}$ $\frac{4}{12}$ $\frac{5}{12}$ $\frac{6}{12}$ $\frac{7}{12}$ $\frac{8}{12}$ $\frac{9}{12}$ $\frac{10}{12}$ $\frac{11}{12}$ 1

0 $\frac{1}{16}$ $\frac{2}{16}$ $\frac{3}{16}$ $\frac{4}{16}$ $\frac{5}{16}$ $\frac{6}{16}$ $\frac{7}{16}$ $\frac{8}{16}$ $\frac{9}{16}$ $\frac{10}{16}$ $\frac{11}{16}$ $\frac{12}{16}$ $\frac{13}{16}$ $\frac{14}{16}$ $\frac{15}{16}$ 1

$\frac{0}{3}$ $\frac{1}{3}$ $\frac{2}{3}$ $\frac{3}{3}$

0 $\frac{1}{2}$ 1

$\frac{0}{4}$ $\frac{1}{4}$ $\frac{2}{4}$ $\frac{3}{4}$ $\frac{4}{4}$

0 $\frac{1}{2}$ 1

$\frac{0}{5}$ $\frac{1}{5}$ $\frac{2}{5}$ $\frac{3}{5}$ $\frac{4}{5}$ $\frac{5}{5}$

0 $\frac{1}{2}$ 1

$\frac{0}{6}$ $\frac{1}{6}$ $\frac{2}{6}$ $\frac{3}{6}$ $\frac{4}{6}$ $\frac{5}{6}$ $\frac{6}{6}$

0 $\frac{1}{2}$ 1

$\frac{0}{8}$ $\frac{1}{8}$ $\frac{2}{8}$ $\frac{3}{8}$ $\frac{4}{8}$ $\frac{5}{8}$ $\frac{6}{8}$ $\frac{7}{8}$ $\frac{8}{8}$

0 $\frac{1}{2}$ 1

$\frac{0}{9}$ $\frac{1}{9}$ $\frac{2}{9}$ $\frac{3}{9}$ $\frac{4}{9}$ $\frac{5}{9}$ $\frac{6}{9}$ $\frac{7}{9}$ $\frac{8}{9}$ $\frac{9}{9}$

0 $\frac{1}{2}$ 1

$\frac{0}{10}$ $\frac{1}{10}$ $\frac{2}{10}$ $\frac{3}{10}$ $\frac{4}{10}$ $\frac{5}{10}$ $\frac{6}{10}$ $\frac{7}{10}$ $\frac{8}{10}$ $\frac{9}{10}$ $\frac{10}{10}$

0 $\frac{1}{2}$ 1

$\frac{0}{12}$ $\frac{1}{12}$ $\frac{2}{12}$ $\frac{3}{12}$ $\frac{4}{12}$ $\frac{5}{12}$ $\frac{6}{12}$ $\frac{7}{12}$ $\frac{8}{12}$ $\frac{9}{12}$ $\frac{10}{12}$ $\frac{11}{12}$ $\frac{12}{12}$

0 $\frac{1}{2}$ 1

$\frac{0}{3}$ $\frac{1}{3}$ $\frac{2}{3}$ $\frac{3}{3}$ $\frac{4}{3}$ $\frac{5}{3}$ $\frac{6}{3}$ $\frac{7}{3}$ $\frac{8}{3}$ $\frac{9}{3}$

0 $\frac{1}{2}$ 1 $1\frac{1}{2}$ 2 $2\frac{1}{2}$ 3

$\frac{0}{4}$ $\frac{1}{4}$ $\frac{2}{4}$ $\frac{3}{4}$ $\frac{4}{4}$ $\frac{5}{4}$ $\frac{6}{4}$ $\frac{7}{4}$ $\frac{8}{4}$ $\frac{9}{4}$ $\frac{10}{4}$ $\frac{11}{4}$ $\frac{12}{4}$

0 $\frac{1}{2}$ 1 $1\frac{1}{2}$ 2 $2\frac{1}{2}$ 3

$\frac{0}{5}$ $\frac{1}{5}$ $\frac{2}{5}$ $\frac{3}{5}$ $\frac{4}{5}$ $\frac{5}{5}$ $\frac{6}{5}$ $\frac{7}{5}$ $\frac{8}{5}$ $\frac{9}{5}$ $\frac{10}{5}$ $\frac{11}{5}$ $\frac{12}{5}$ $\frac{13}{5}$ $\frac{14}{5}$ $\frac{15}{5}$

0 $\frac{1}{2}$ 1 $1\frac{1}{2}$ 2 $2\frac{1}{2}$ 3

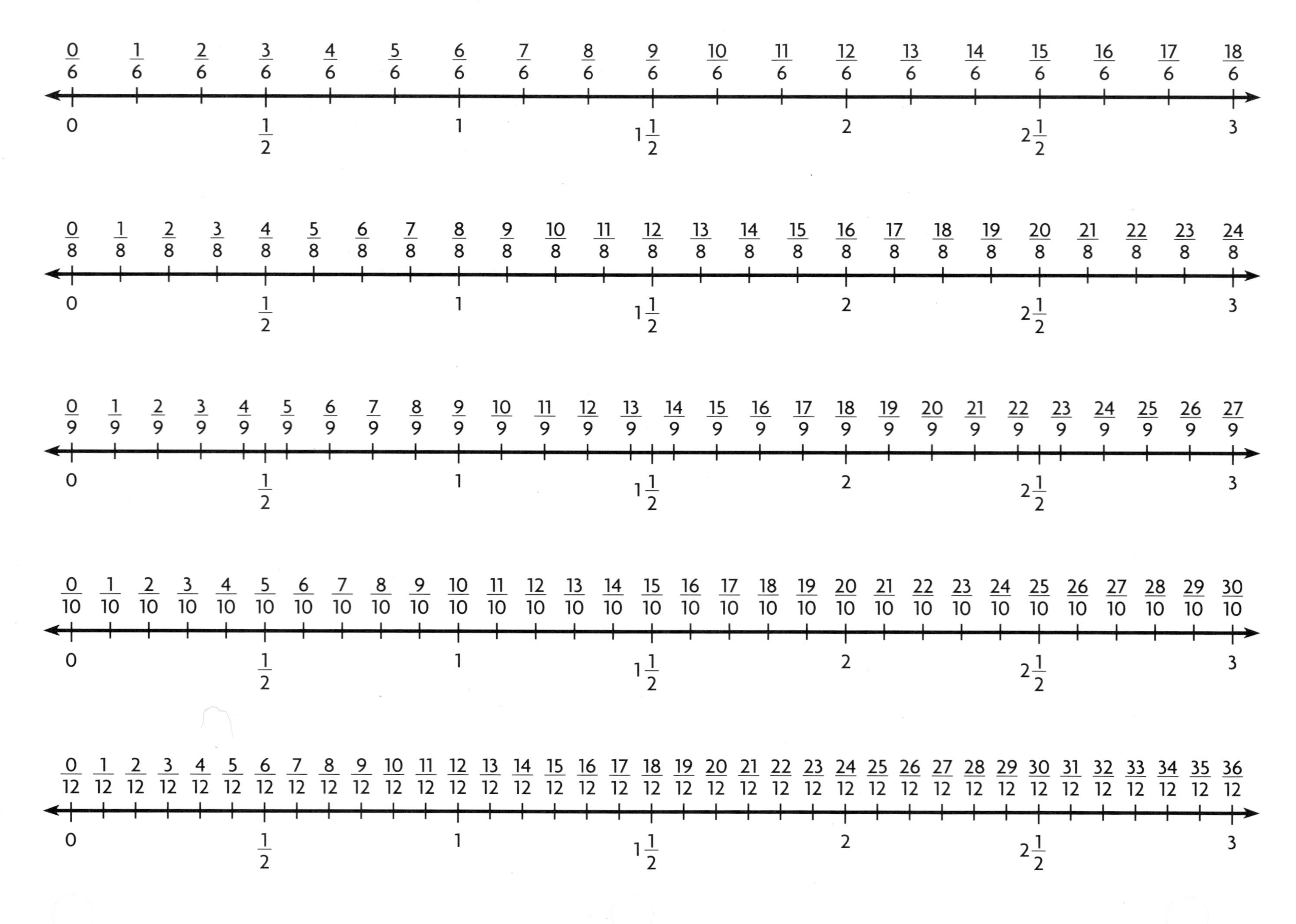

0/6 1/6 2/6 3/6 4/6 5/6 6/6 7/6 8/6 9/6 10/6 11/6 12/6 13/6 14/6 15/6 16/6 17/6 18/6
0 1/2 1 1 1/2 2 2 1/2 3
0/8 1/8 2/8 3/8 4/8 5/8 6/8 7/8 8/8 9/8 10/8 11/8 12/8 13/8 14/8 15/8 16/8 17/8 18/8 19/8 20/8 21/8 22/8 23/8 24/8
0 1/2 1 1 1/2 2 2 1/2 3
0/9 1/9 2/9 3/9 4/9 5/9 6/9 7/9 8/9 9/9 10/9 11/9 12/9 13/9 14/9 15/9 16/9 17/9 18/9 19/9 20/9 21/9 22/9 23/9 24/9 25/9 26/9 27/9
0 1/2 1 1 1/2 2 2 1/2 3
0/10 1/10 2/10 3/10 4/10 5/10 6/10 7/10 8/10 9/10 10/10 11/10 12/10 13/10 14/10 15/10 16/10 17/10 18/10 19/10 20/10 21/10 22/10 23/10 24/10 25/10 26/10 27/10 28/10 29/10 30/10
0 1/2 1 1 1/2 2 2 1/2 3
0/12 1/12 2/12 3/12 4/12 5/12 6/12 7/12 8/12 9/12 10/12 11/12 12/12 13/12 14/12 15/12 16/12 17/12 18/12 19/12 20/12 21/12 22/12 23/12 24/12 25/12 26/12 27/12 28/12 29/12 30/12 31/12 32/12 33/12 34/12 35/12 36/12
0 1/2 1 1 1/2 2 2 1/2 3

© Harcourt

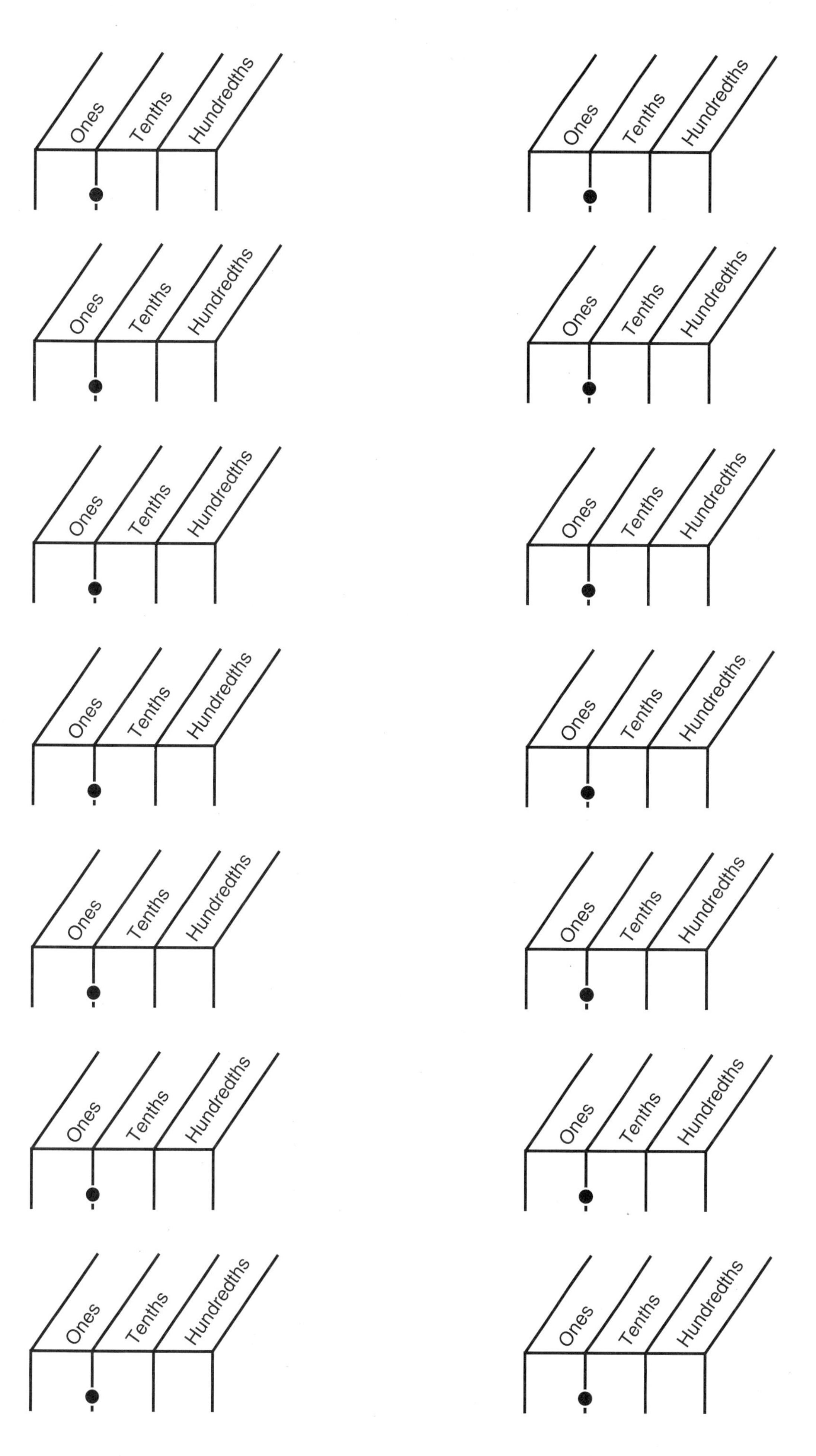
Ones
Tenths
Hundredths

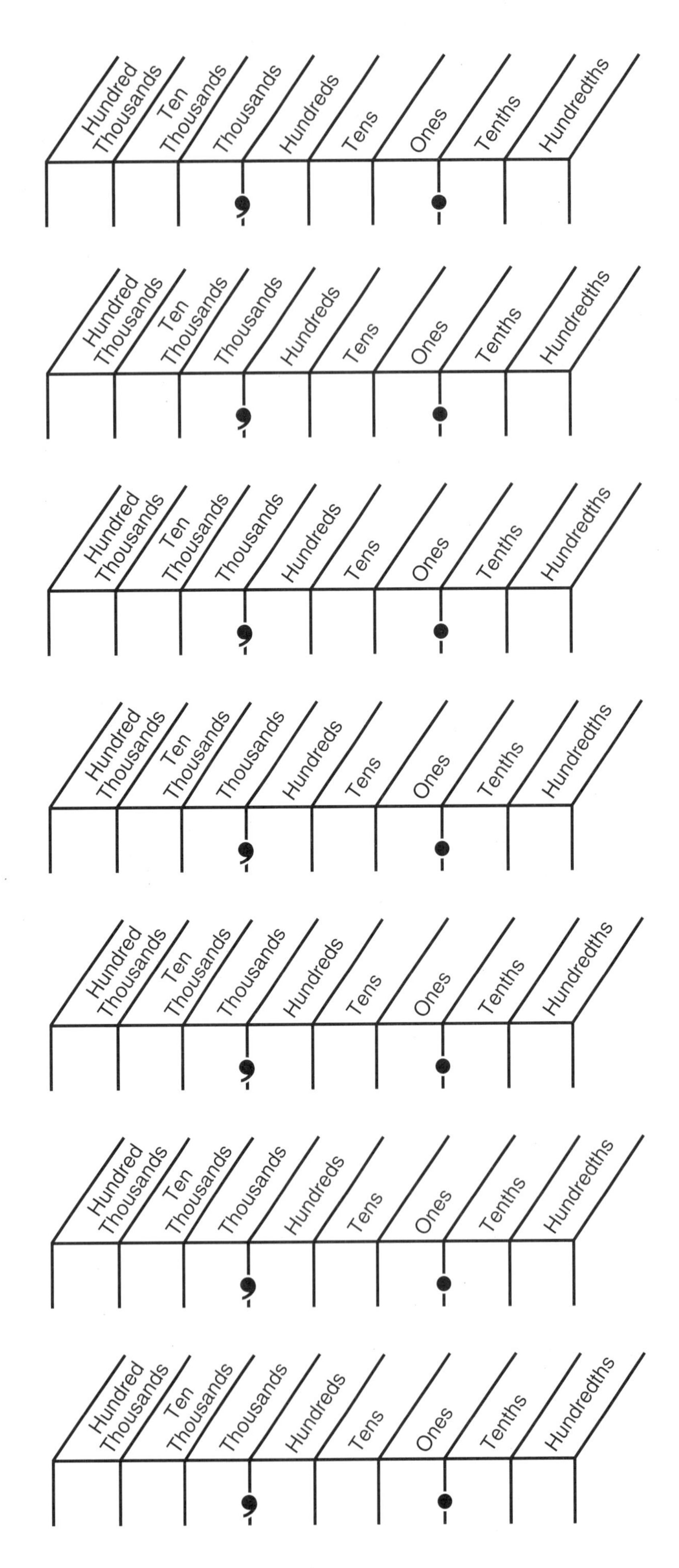
Hundred Thousands
Ten Thousands
Thousands
Hundreds
Tens
Ones
Tenths
Hundredths
Hundred Thousands
Ten Thousands
Thousands
Hundreds
Tens
Ones
Tenths
Hundredths
Hundred Thousands
Ten Thousands
Thousands
Hundreds
Tens
Ones
Tenths
Hundredths
Hundred Thousands
Ten Thousands
Thousands
Hundreds
Tens
Ones
Tenths
Hundredths
Hundred Thousands
Ten Thousands
Thousands
Hundreds
Tens
Ones
Tenths
Hundredths
Hundred Thousands
Ten Thousands
Thousands
Hundreds
Tens
Ones
Tenths
Hundredths
Hundred Thousands
Ten Thousands
Thousands
Hundreds
Tens
Ones
Tenths
Hundredths

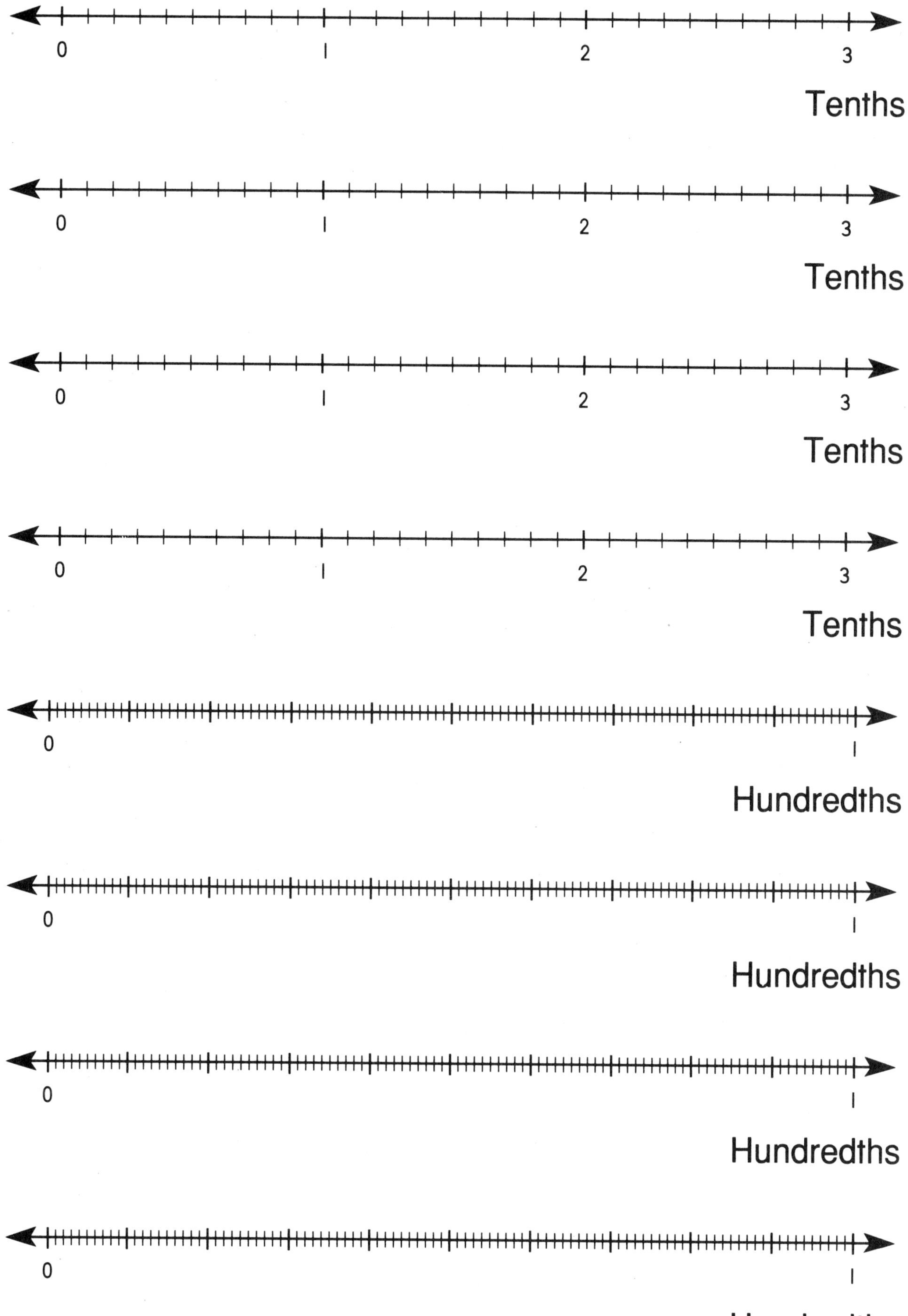
0
1
2
3
Tenths
0
1
2
3
Tenths
0
1
2
3
Tenths
0
1
2
3
Tenths
0
1
Hundredths
0
1
Hundredths
0
1
Hundredths
0
1
Hundredths

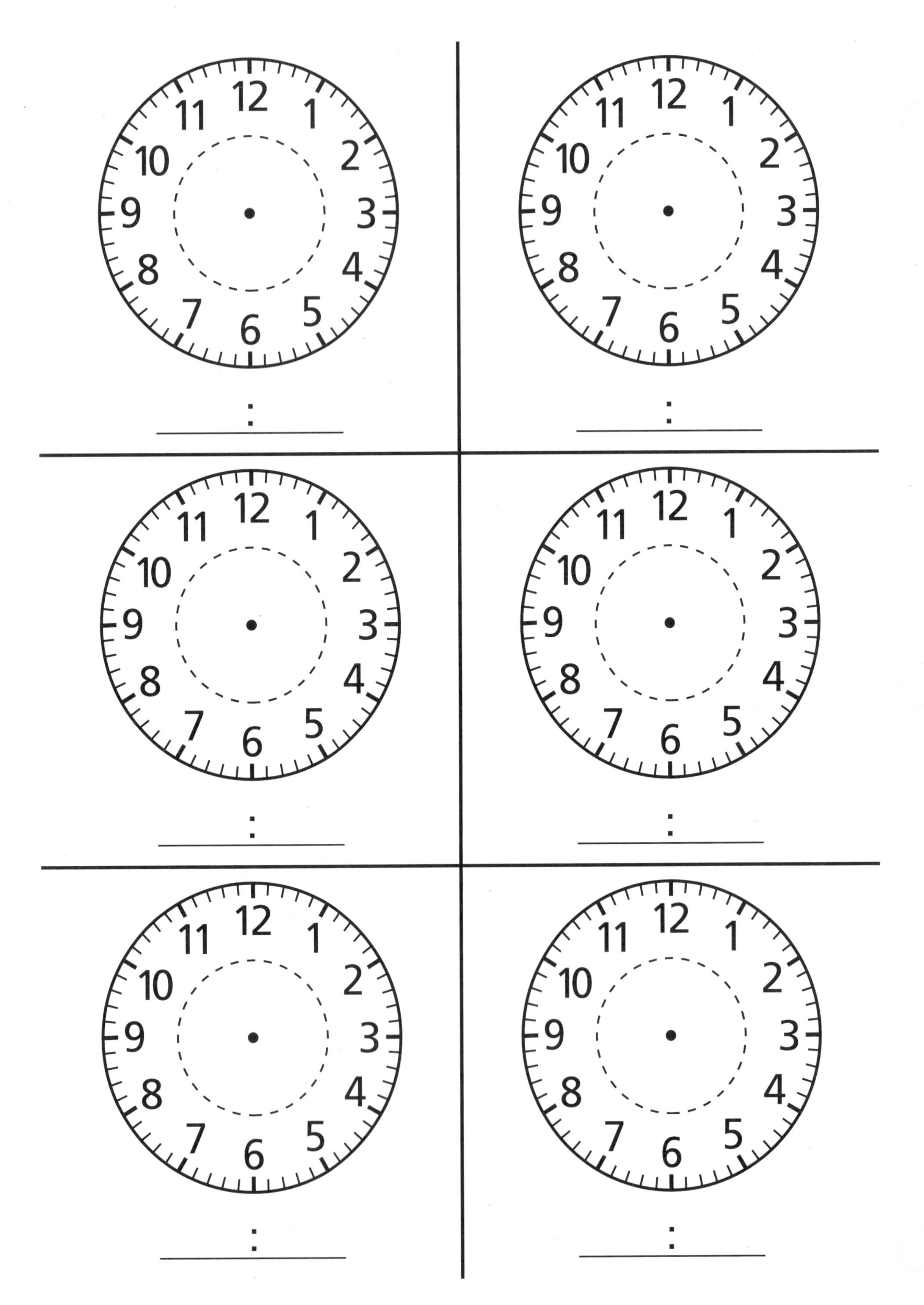
12
1
2
3
4
5
6
7
8
9
10
11
:
12
1
2
3
4
5
6
7
8
9
10
11
:
12
1
2
3
4
5
6
7
8
9
10
11
:
12
1
2
3
4
5
6
7
8
9
10
11
:
12
1
2
3
4
5
6
7
8
9
10
11
:
12
1
2
3
4
5
6
7
8
9
10
11
:

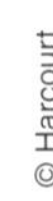

IN GOD WE TRUST
LIBERTY
1999
UNITED STATES OF AMERICA
E·PLURIBUS UNUM
ONE CENT
MONTICELLO
FIVE CENTS
ONE DIME
QUARTER DOLLAR

THE UNITED STATES OF AMERICA
IN GOD WE TRUST
ONE
ONE DOLLAR

THE UNITED STATES OF AMERICA
IN GOD WE TRUST
ONE
ONE DOLLAR

THE UNITED STATES OF AMERICA
IN GOD WE TRUST
ONE
ONE DOLLAR

THE UNITED STATES OF AMERICA
IN GOD WE TRUST
ONE
ONE DOLLAR

FEDERAL RESERVE NOTE
THE UNITED STATES OF AMERICA
BZ 00028071 A
FIVE DOLLARS

FEDERAL RESERVE NOTE
THE UNITED STATES OF AMERICA
BZ 00028071 A
FIVE DOLLARS

FEDERAL RESERVE NOTE
THE UNITED STATES OF AMERICA
BZ 00028071 A
FIVE DOLLARS

FEDERAL RESERVE NOTE
THE UNITED STATES OF AMERICA
BZ 00028071 A
FIVE DOLLARS

THE UNITED STATES OF AMERICA
5
FIVE DOLLARS

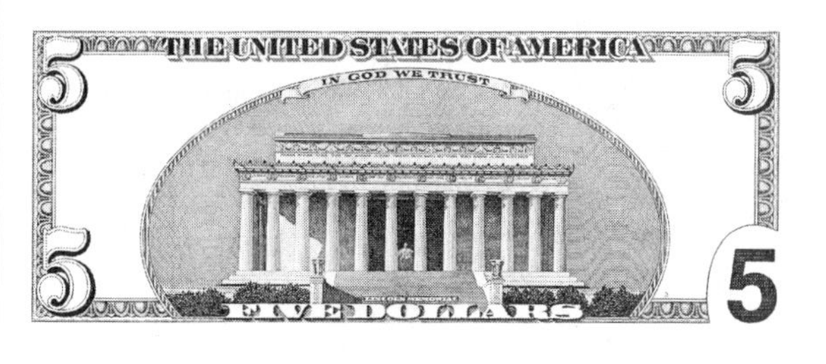
THE UNITED STATES OF AMERICA
5
FIVE DOLLARS

THE UNITED STATES OF AMERICA
IN GOD WE TRUST
FIVE DOLLARS
5

THE UNITED STATES OF AMERICA
IN GOD WE TRUST
FIVE DOLLARS
5

10
FEDERAL RESERVE NOTE
BZ 00008112 A
THE UNITED STATES OF AMERICA
TEN
BZ 00008112 A
TEN DOLLARS

10
FEDERAL RESERVE NOTE
BZ 00008112 A
THE UNITED STATES OF AMERICA
TEN
BZ 00008112 A
TEN DOLLARS

10
FEDERAL RESERVE NOTE
BZ 00008112 A
THE UNITED STATES OF AMERICA
TEN
BZ 00008112 A
TEN DOLLARS

10
FEDERAL RESERVE NOTE
BZ 00008112 A
THE UNITED STATES OF AMERICA
TEN
BZ 00008112 A
TEN DOLLARS

THE UNITED STATES OF AMERICA
IN GOD WE TRUST
TEN DOLLARS
10

THE UNITED STATES OF AMERICA
IN GOD WE TRUST
TEN DOLLARS
10

THE UNITED STATES OF AMERICA
IN GOD WE TRUST
TEN DOLLARS
10

THE UNITED STATES OF AMERICA
IN GOD WE TRUST
TEN DOLLARS
10

FEDERAL RESERVE NOTE
AZ 00833702 A
THE UNITED STATES OF AMERICA
AZ 00833702 A
TWENTY DOLLARS

FEDERAL RESERVE NOTE
AZ 00833702 A
THE UNITED STATES OF AMERICA
AZ 00833702 A
TWENTY DOLLARS

FEDERAL RESERVE NOTE
AZ 00833702 A
THE UNITED STATES OF AMERICA
AZ 00833702 A
TWENTY DOLLARS

FEDERAL RESERVE NOTE
AZ 00833702 A
THE UNITED STATES OF AMERICA
AZ 00833702 A
TWENTY DOLLARS

THE UNITED STATES OF AMERICA
IN GOD WE TRUST
TWENTY DOLLARS
20

THE UNITED STATES OF AMERICA
IN GOD WE TRUST
TWENTY DOLLARS
20

THE UNITED STATES OF AMERICA
IN GOD WE TRUST
TWENTY DOLLARS
20

THE UNITED STATES OF AMERICA
IN GOD WE TRUST
TWENTY DOLLARS
20

FEDERAL RESERVE NOTE
AZ 08588104 A
THE UNITED STATES OF AMERICA
AZ 08588104 A
DOLLARS

THE UNITED STATES OF AMERICA
IN GOD WE TRUST
50

____

Sunday	Monday	Tuesday	Wednesday	Thursday	Friday	Saturday

inches 1 2 3 4 5 6 7 8 9

inches 1 2 3 4 5 6 7 8 9

cm 1 2 3 4 5 6 7 8 9 10 11 12 13 14 15 16 17 18 19 20 21 22

1 dm (decimeter) 2 dm

cm 1 2 3 4 5 6 7 8 9 10 11 12 13 14 15 16 17 18 19 20 21 22

1 dm (decimeter) 2 dm

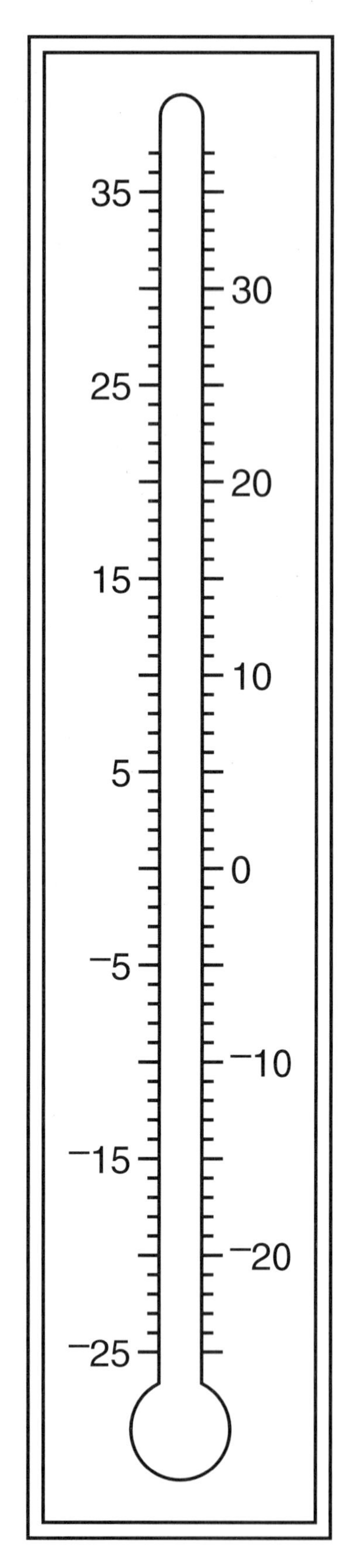

Celsius

____ °C

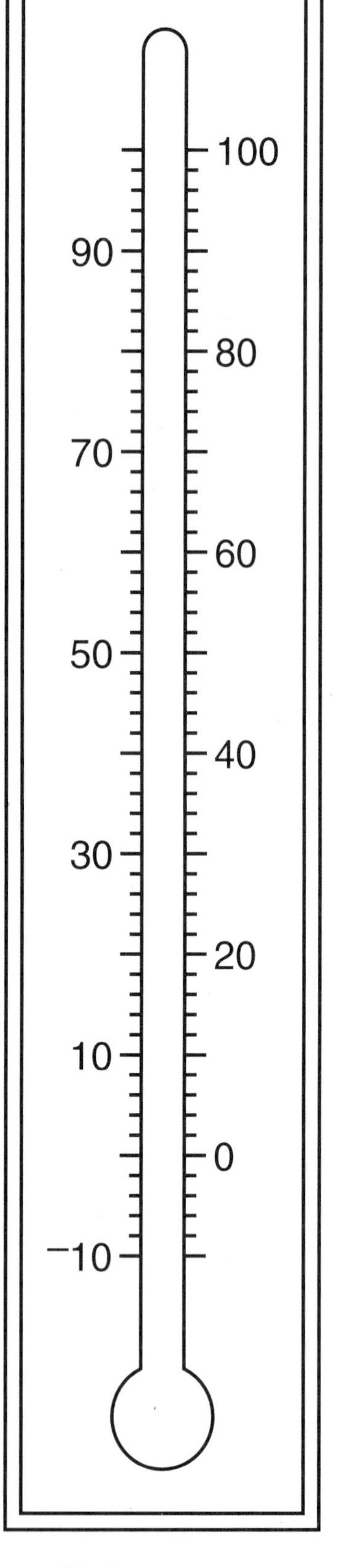

Fahrenheit

____ °F

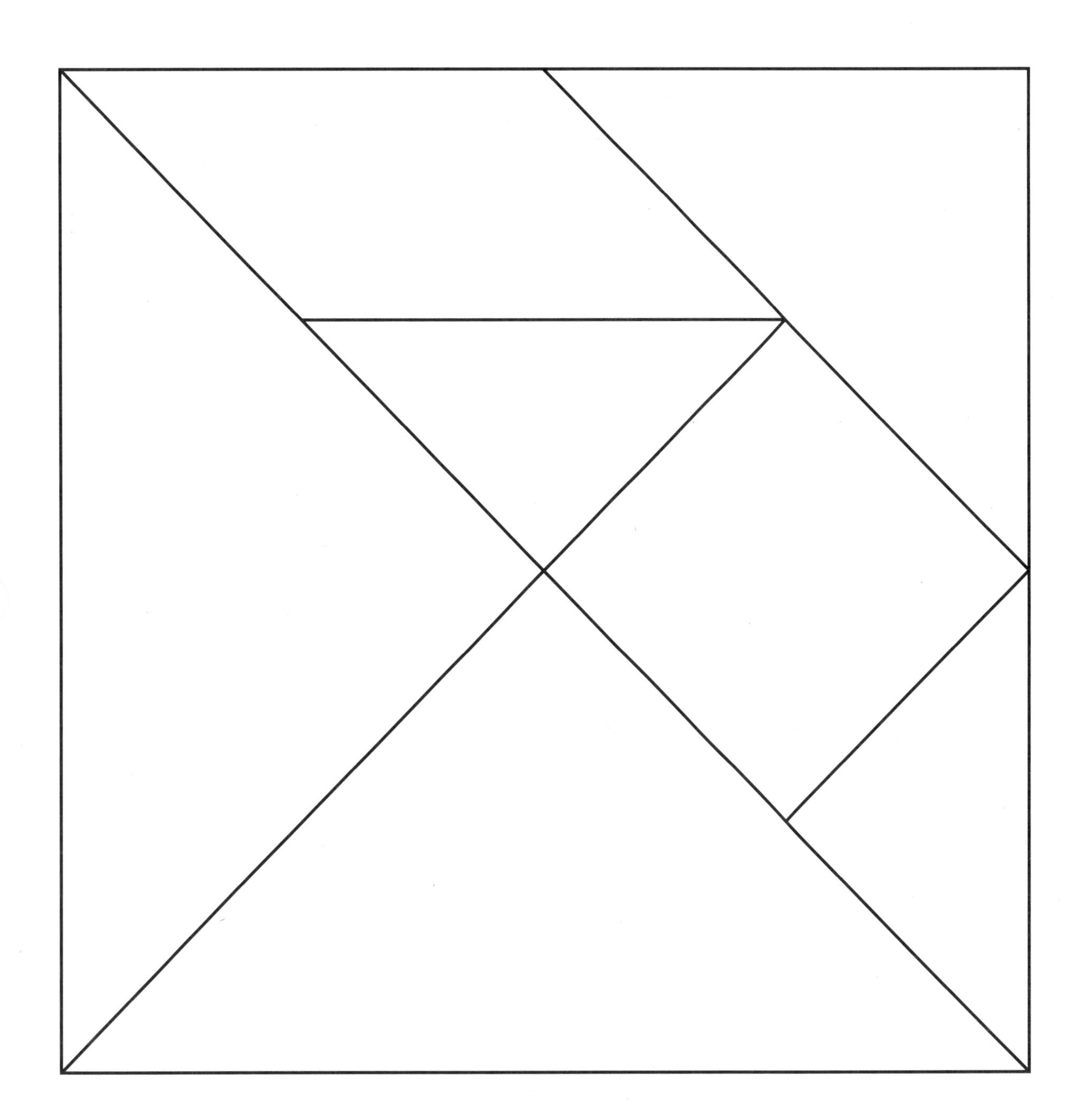

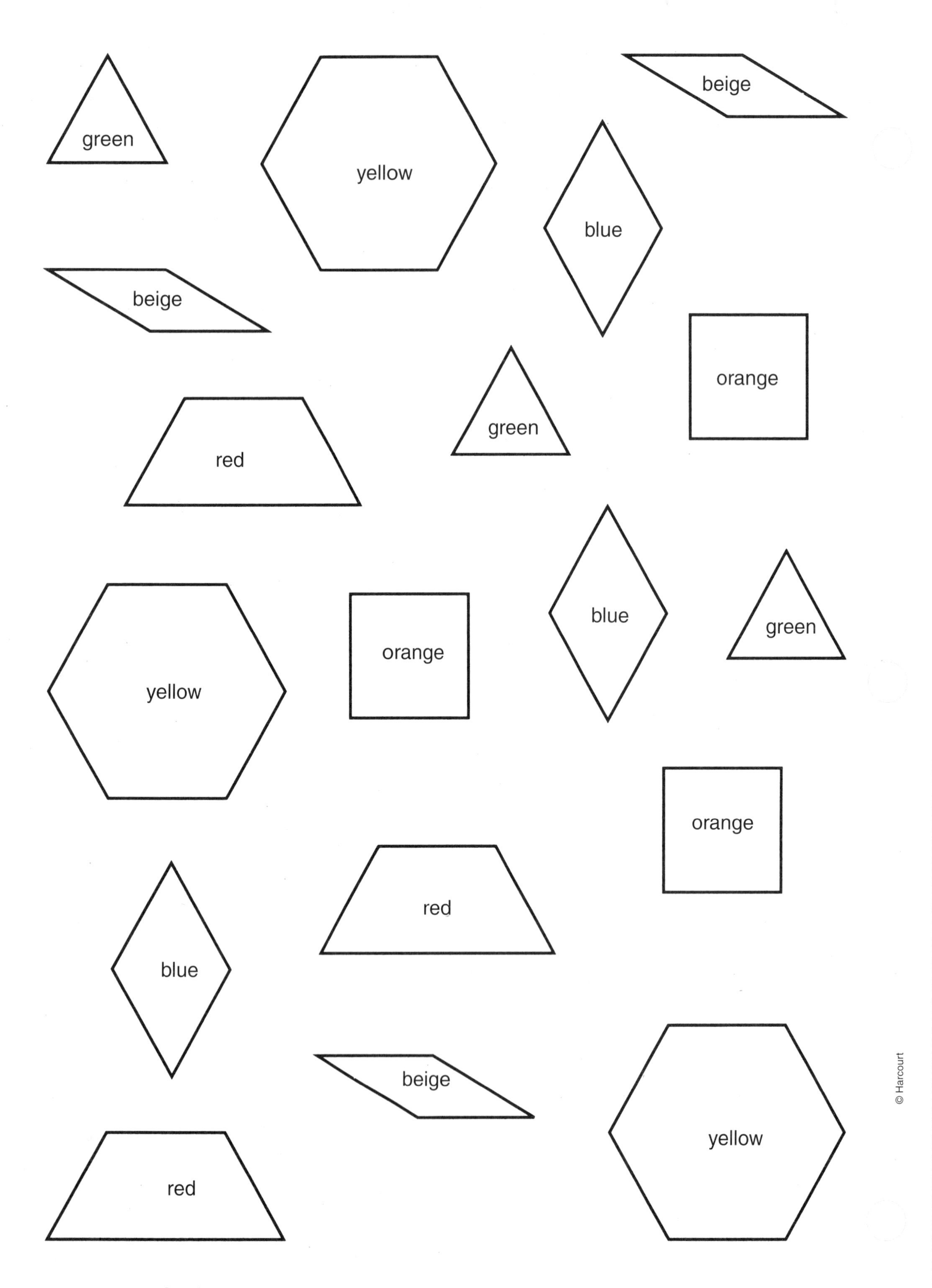
green
yellow
beige
blue
beige
orange
green
red
blue
green
orange
yellow
orange
red
blue
beige
yellow
red

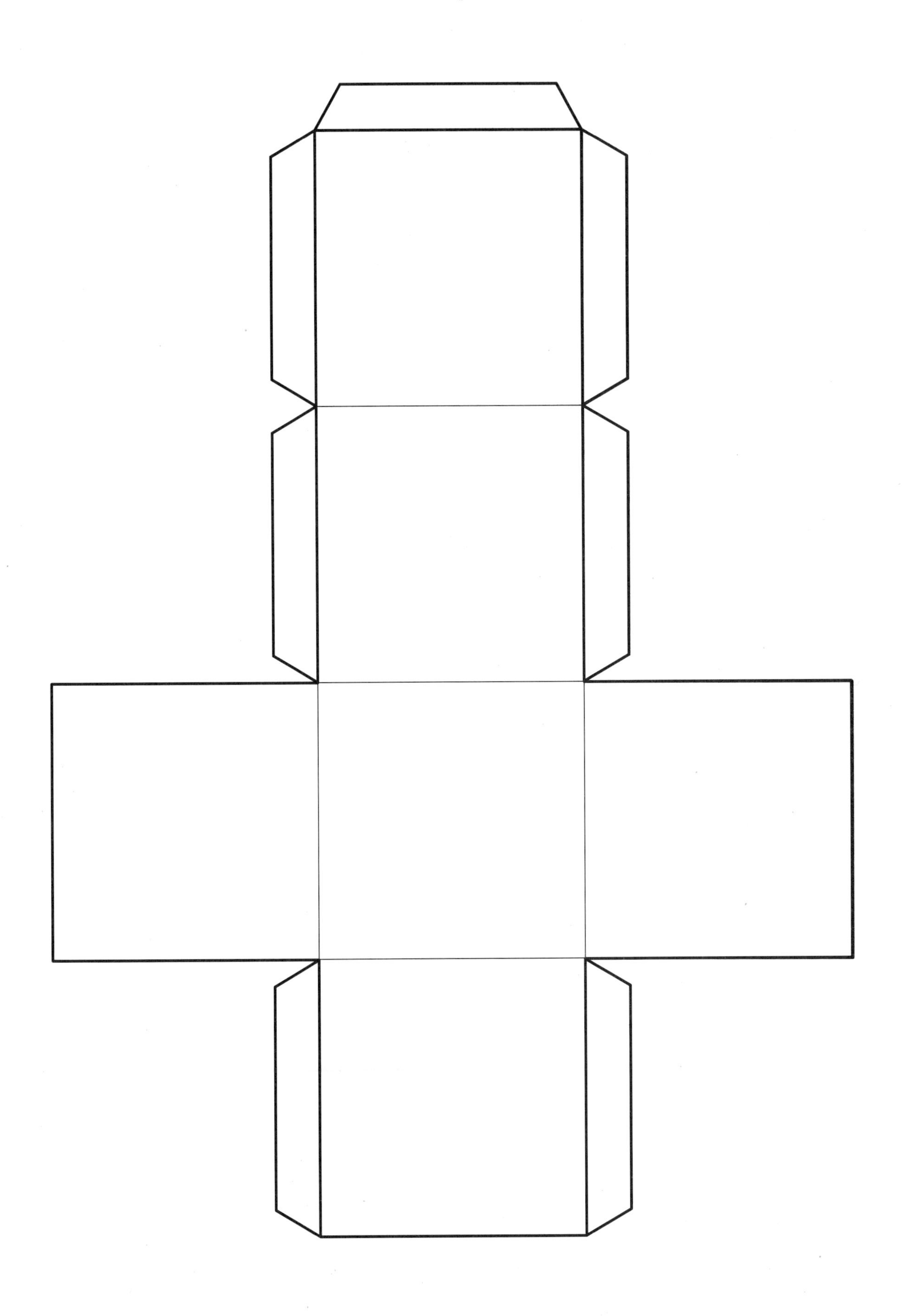

Cube Pattern

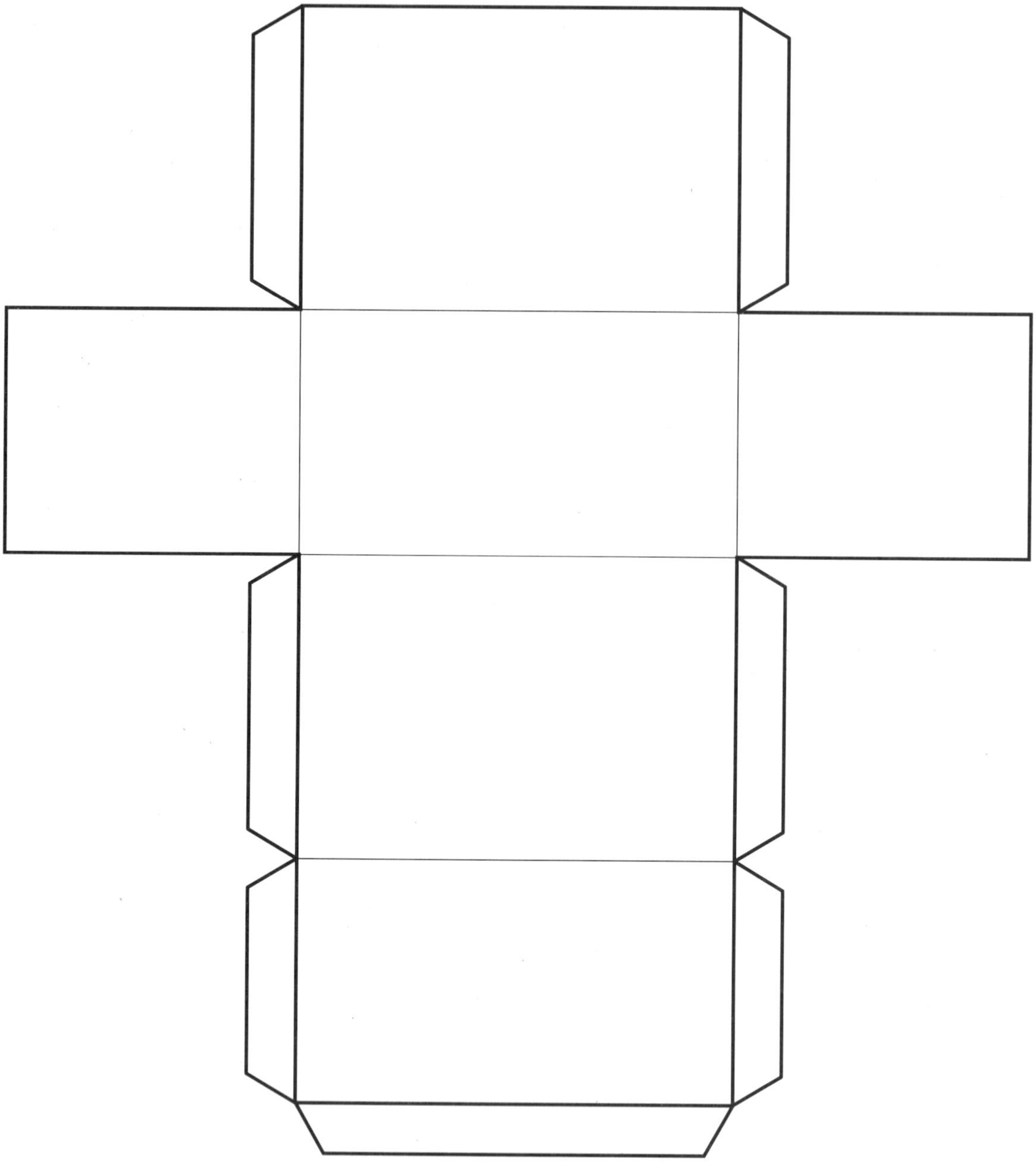

Rectangular Prism Pattern

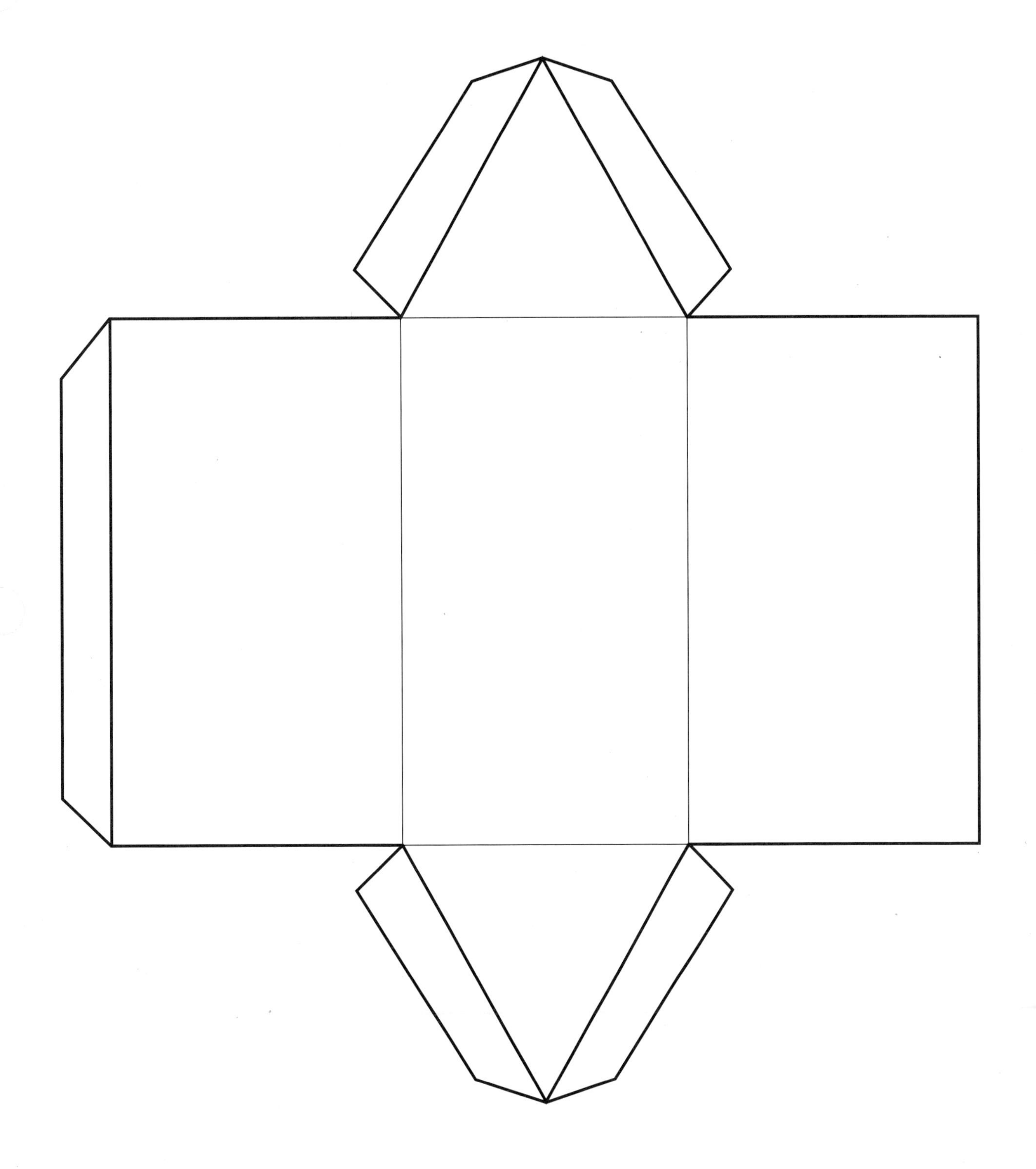

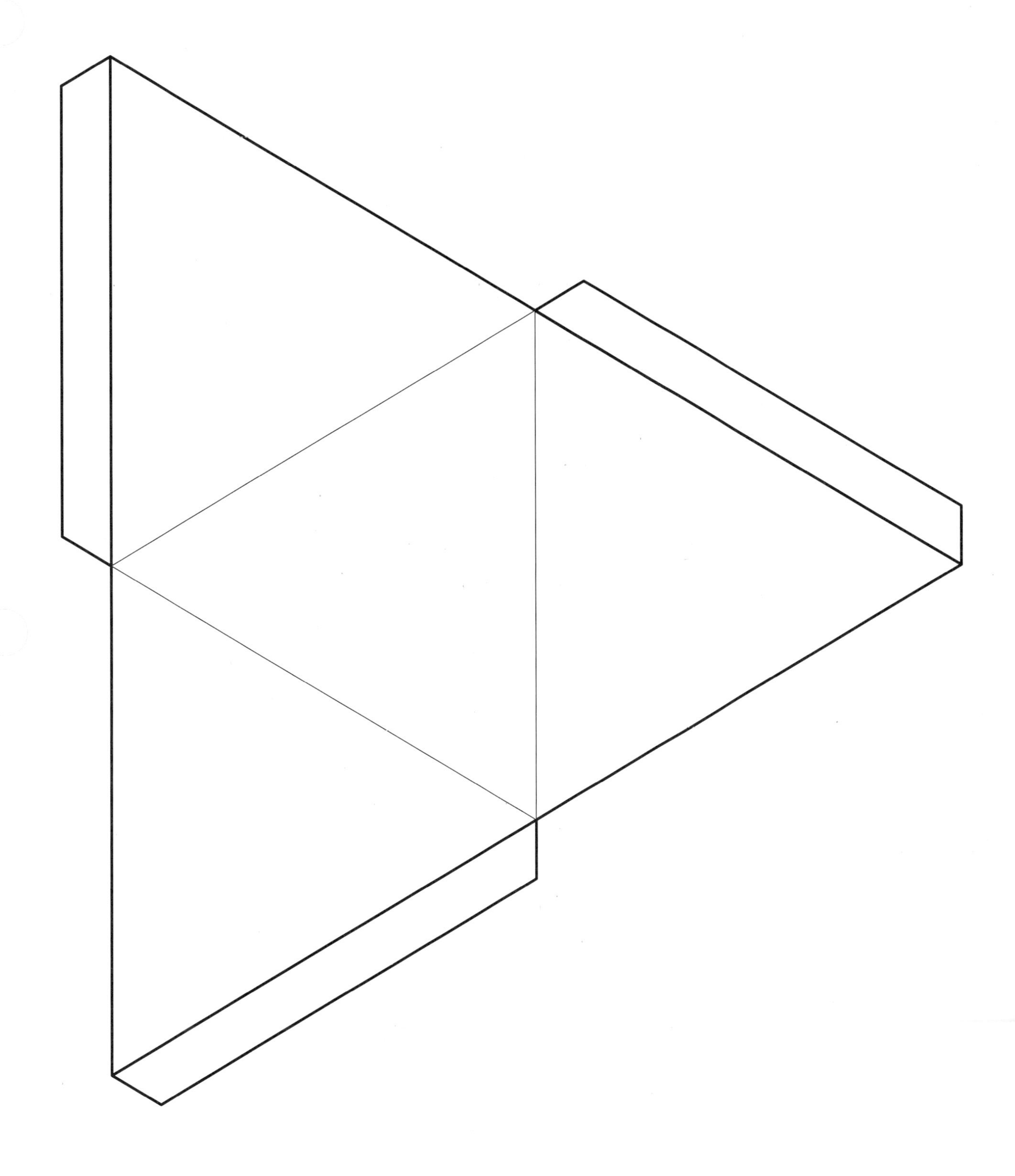

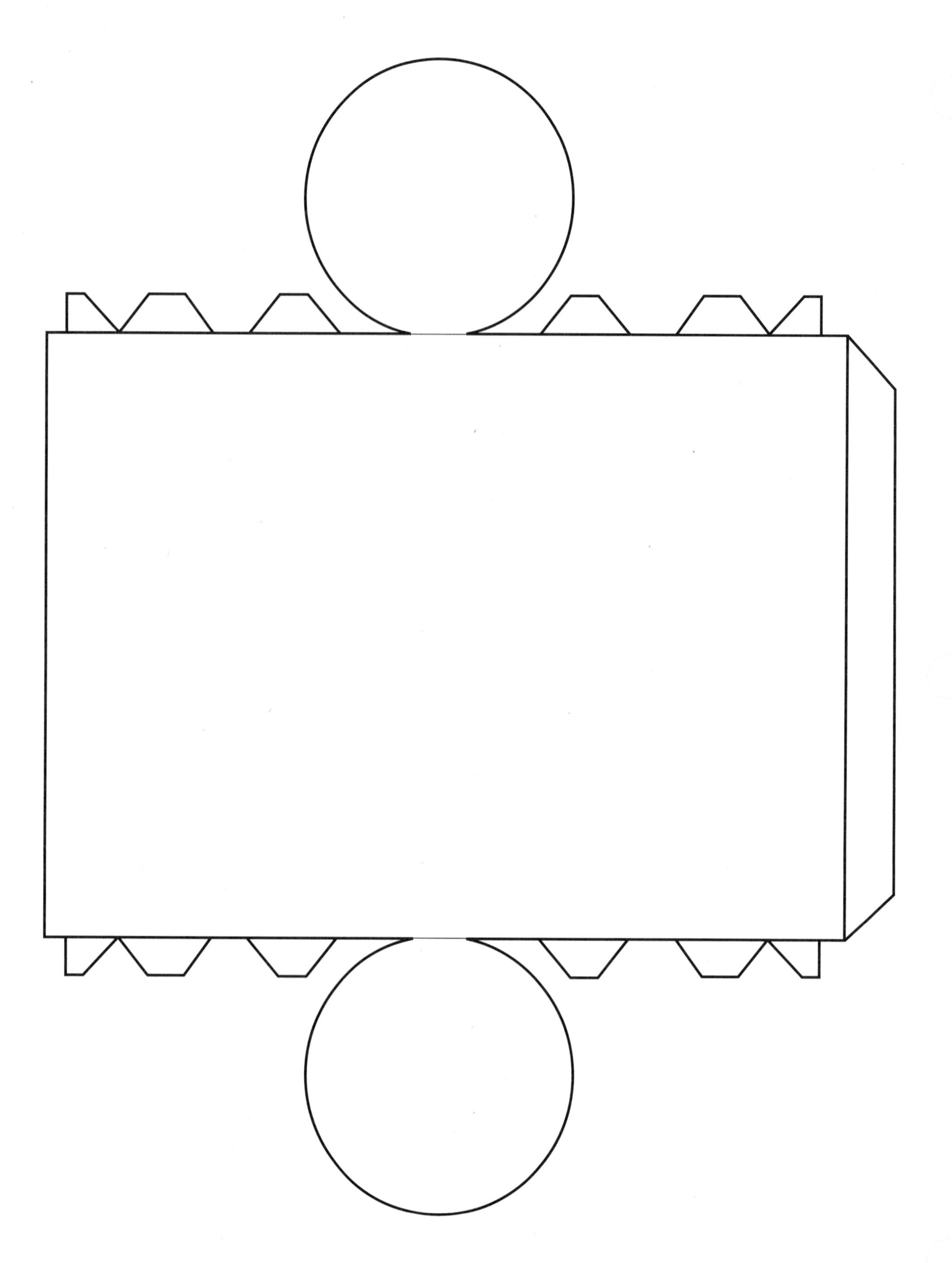

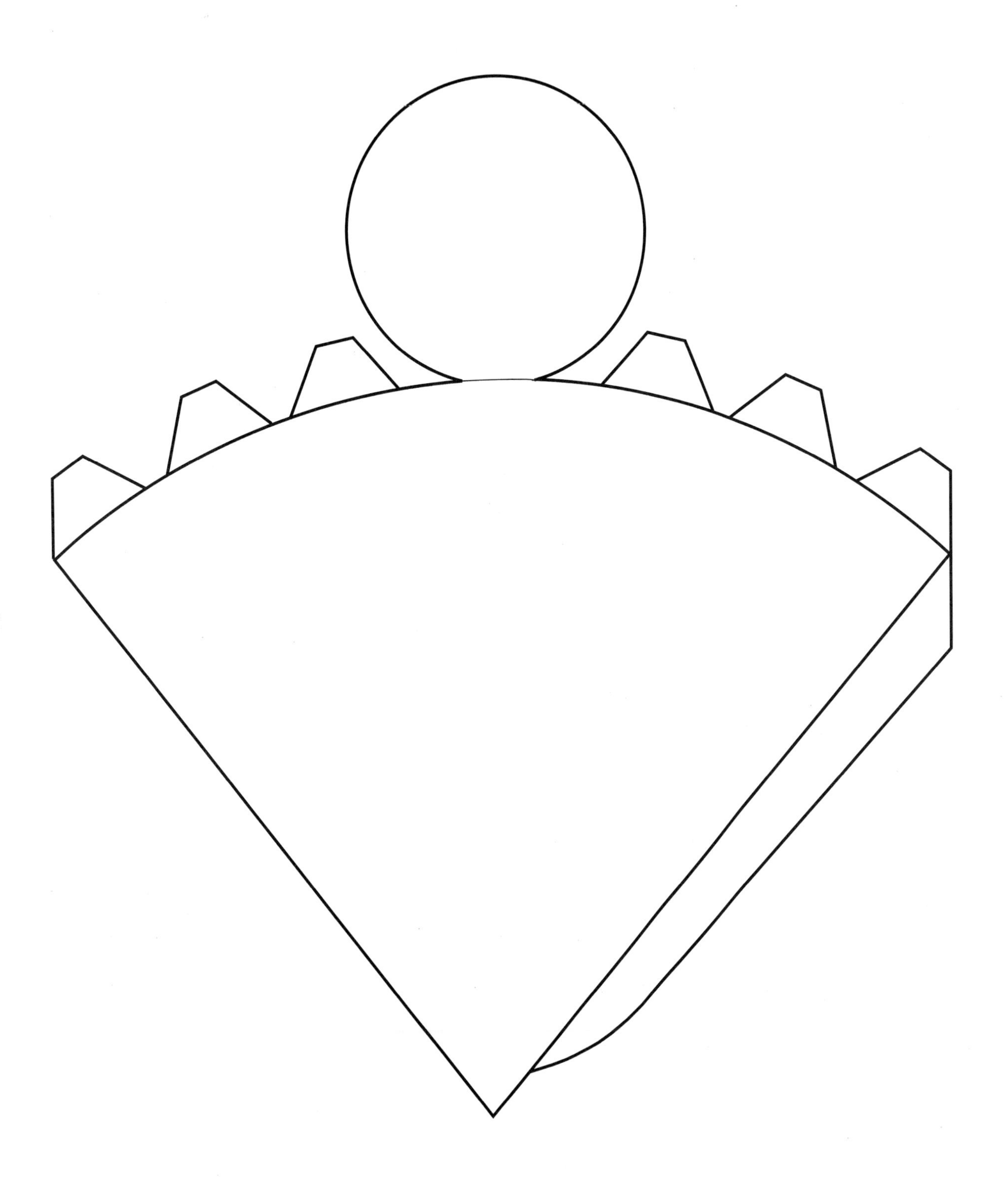

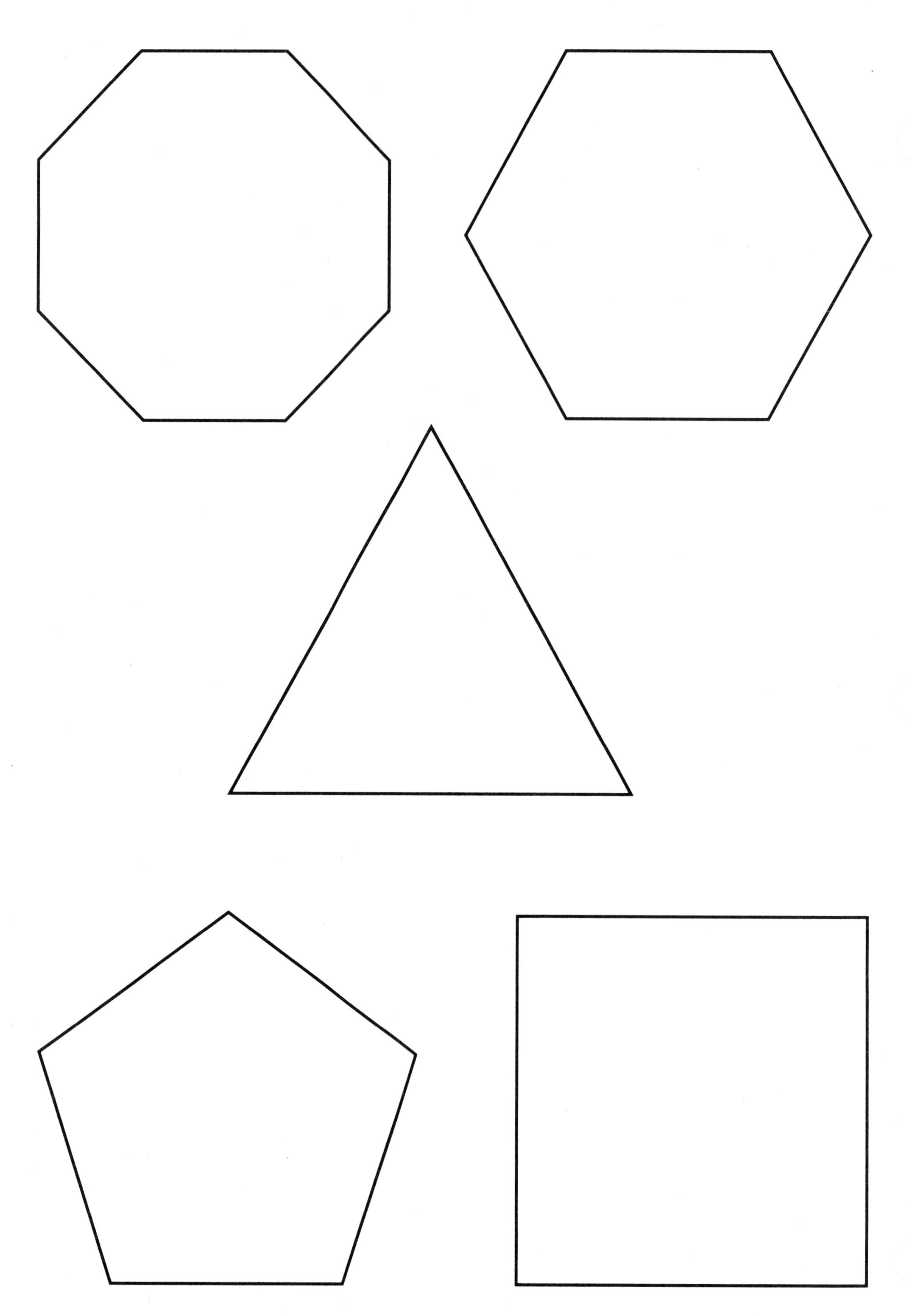

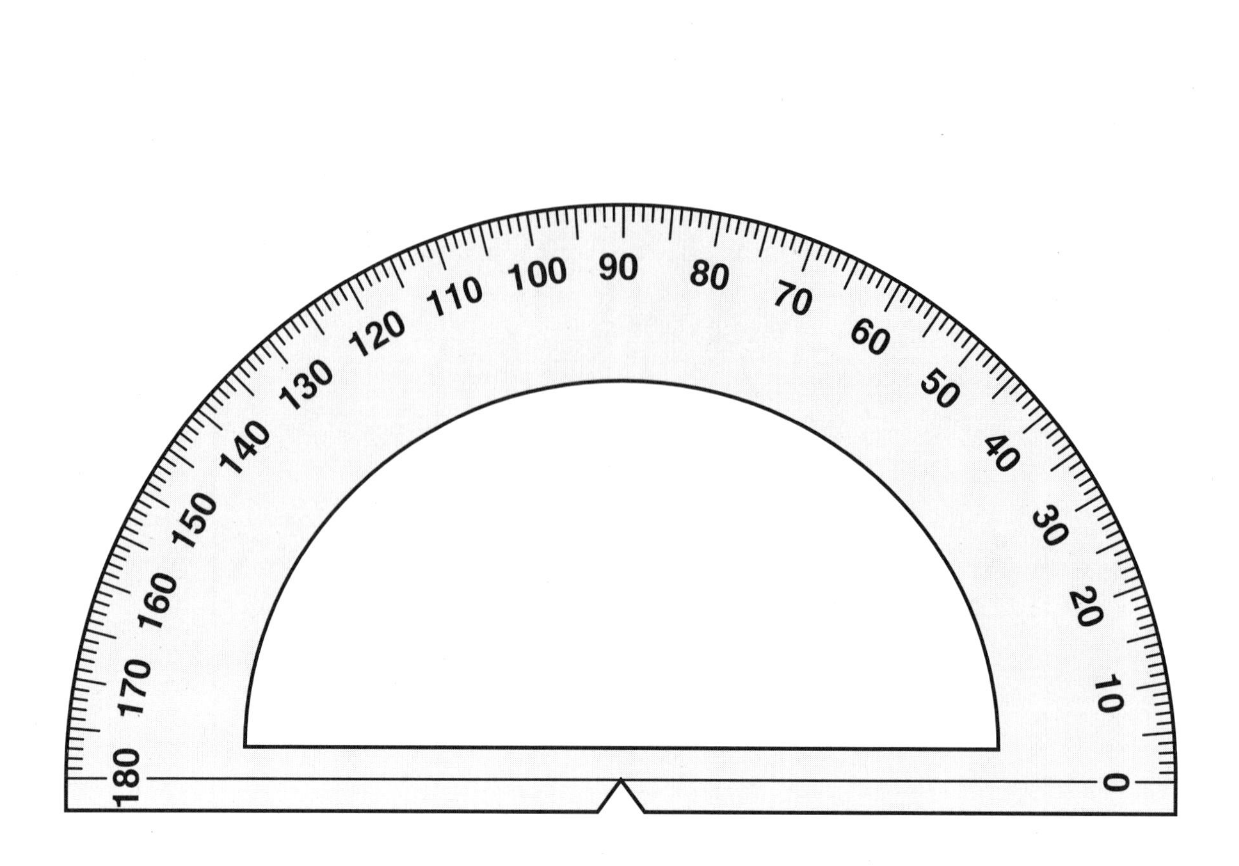
180
170
160
150
140
130
120
110
100
90
80
70
60
50
40
30
20
10
0

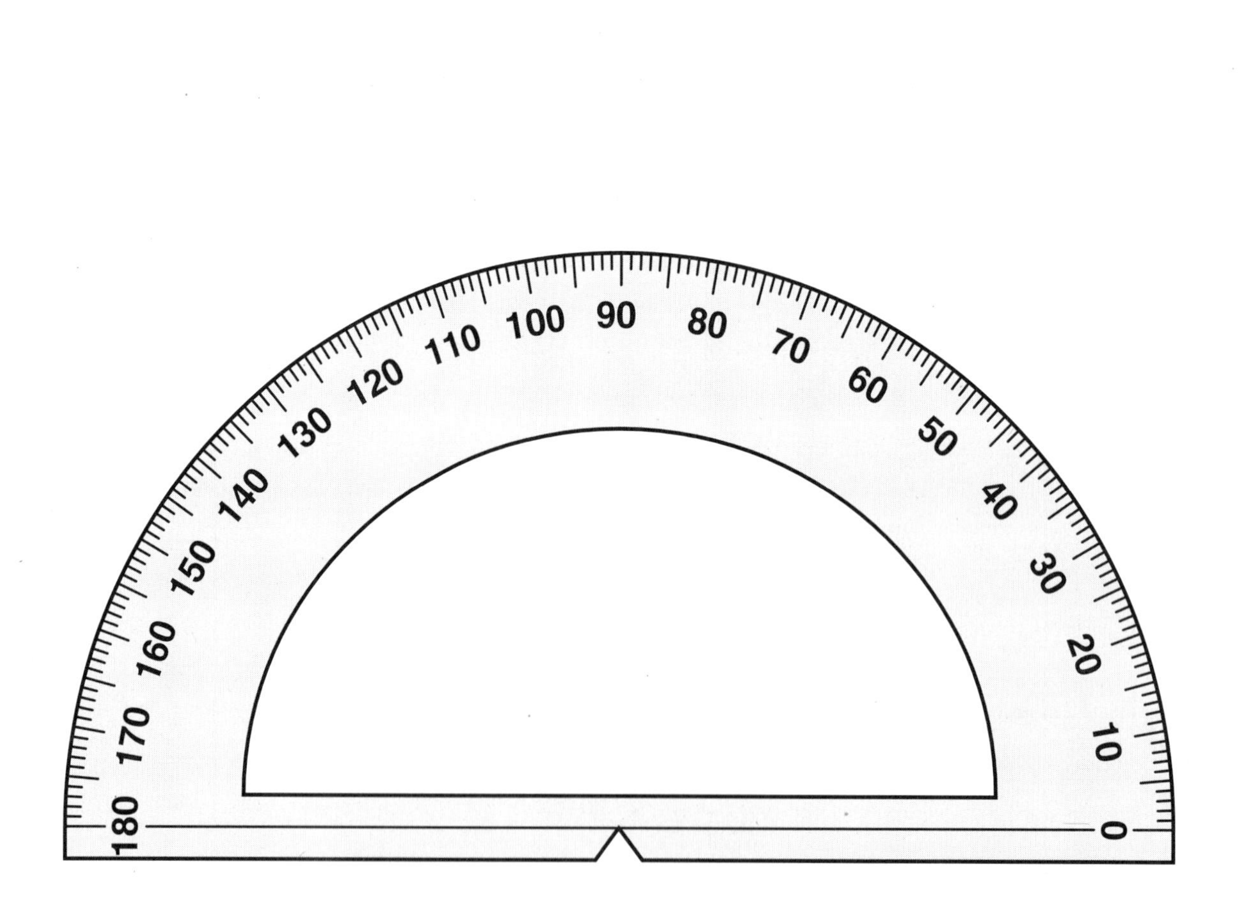
180
170
160
150
140
130
120
110
100
90
80
70
60
50
40
30
20
10
0

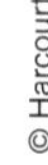
© Harcourt

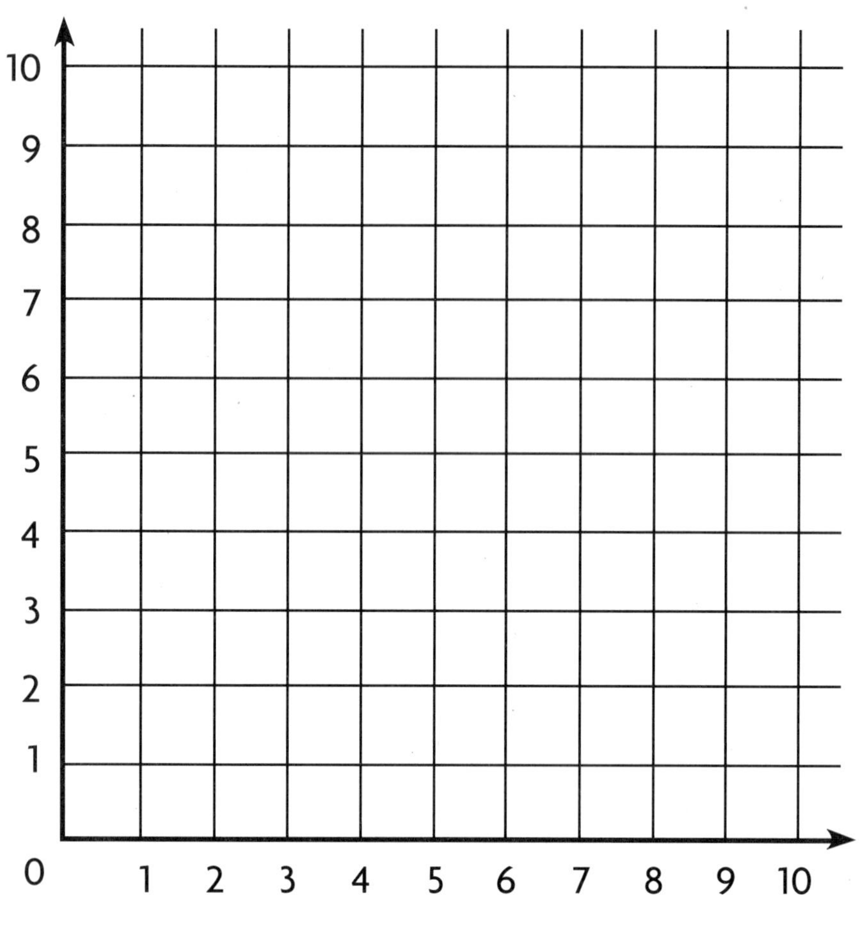
10
9
8
7
6
5
4
3
2
1
0
1 2 3 4 5 6 7 8 9 10

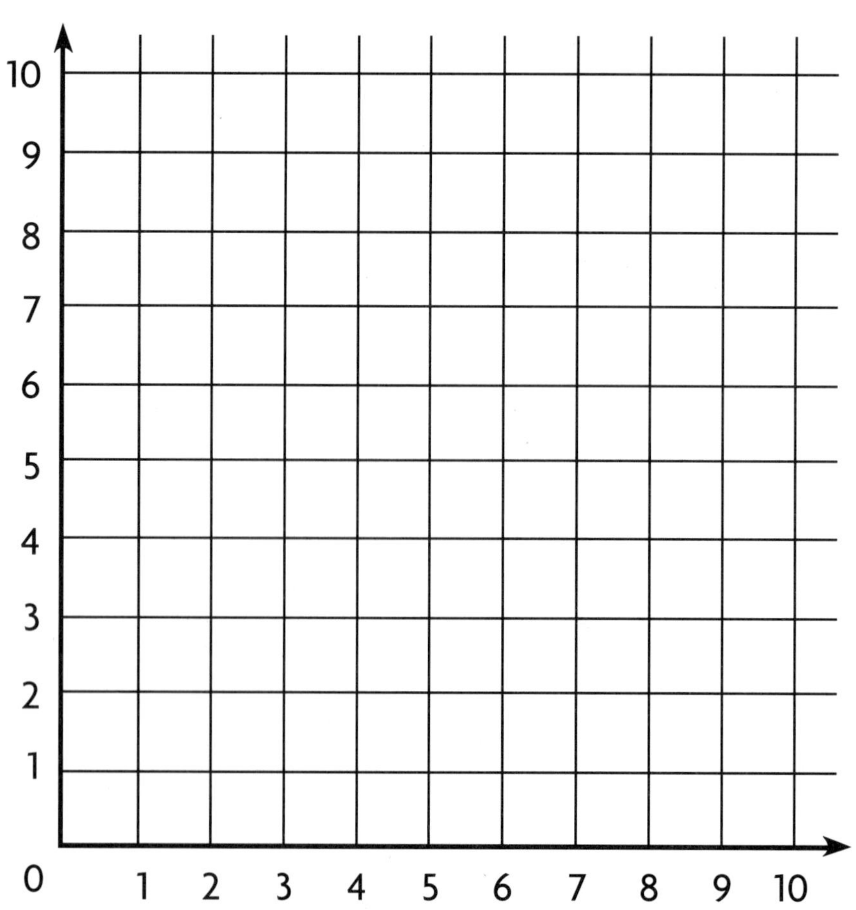
10
9
8
7
6
5
4
3
2
1
0
1 2 3 4 5 6 7 8 9 10

© Harcourt

Title ______________________________

0

____

Title ________________________________

0

Title____________________

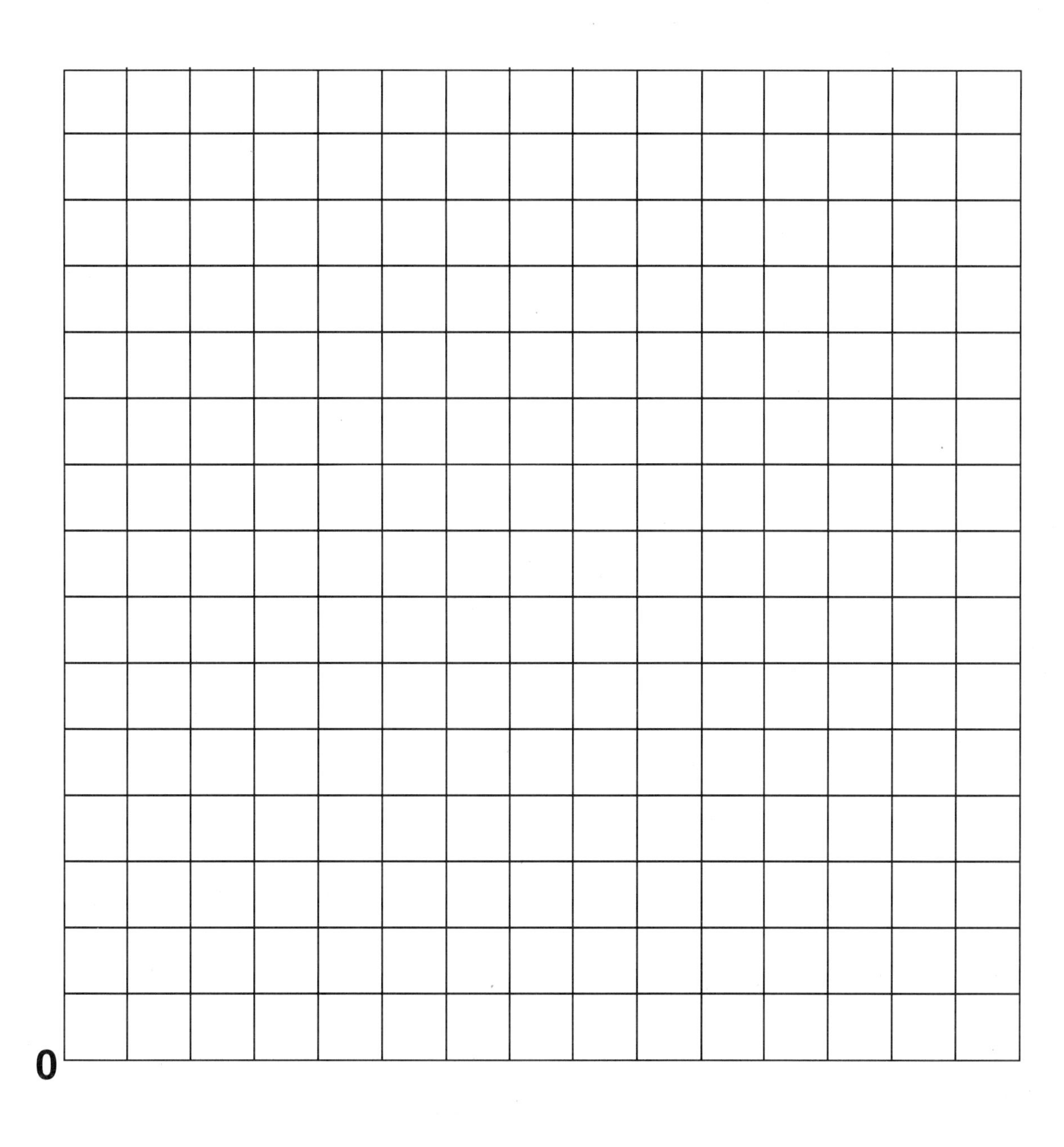

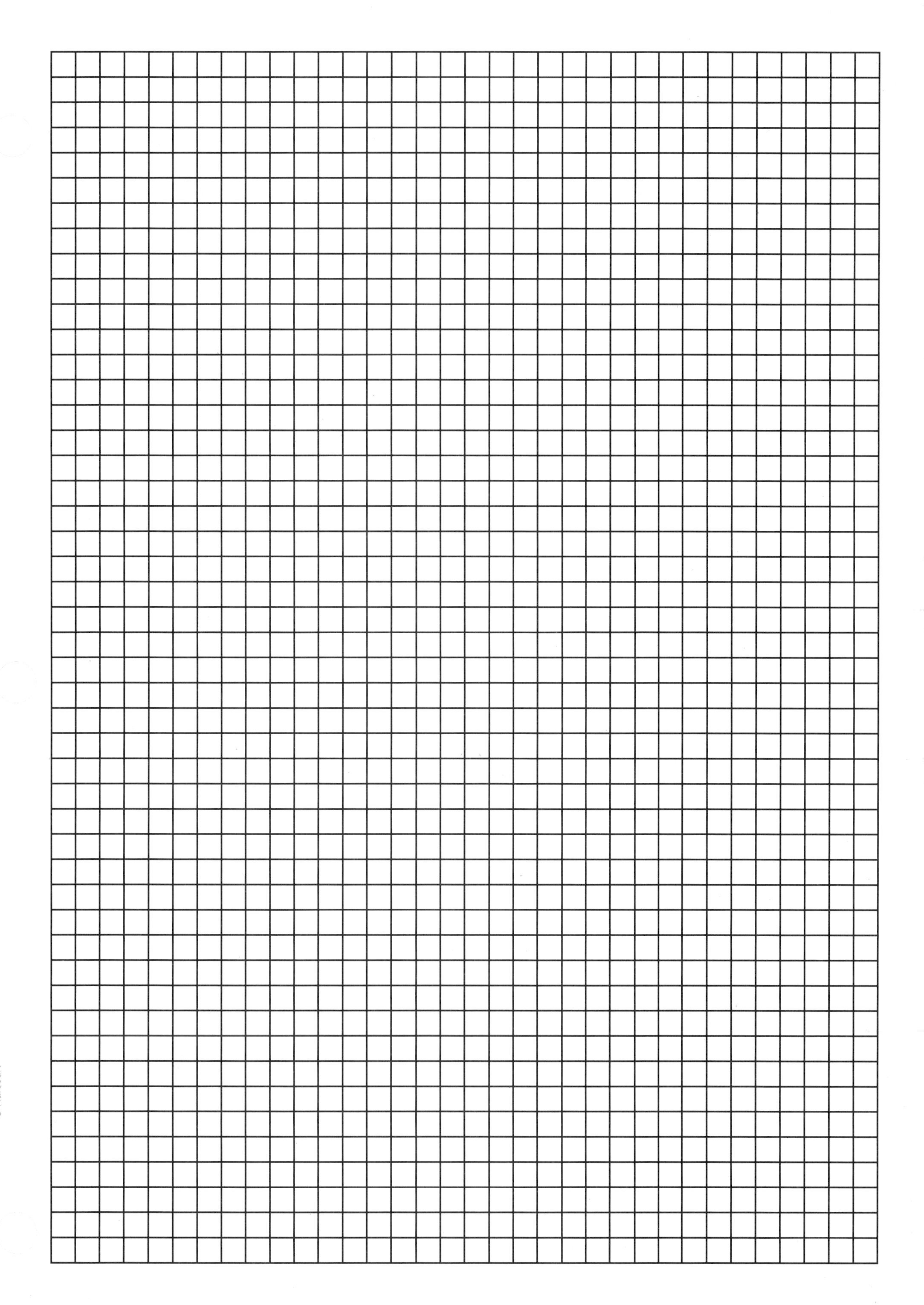

	Tally	Frequency

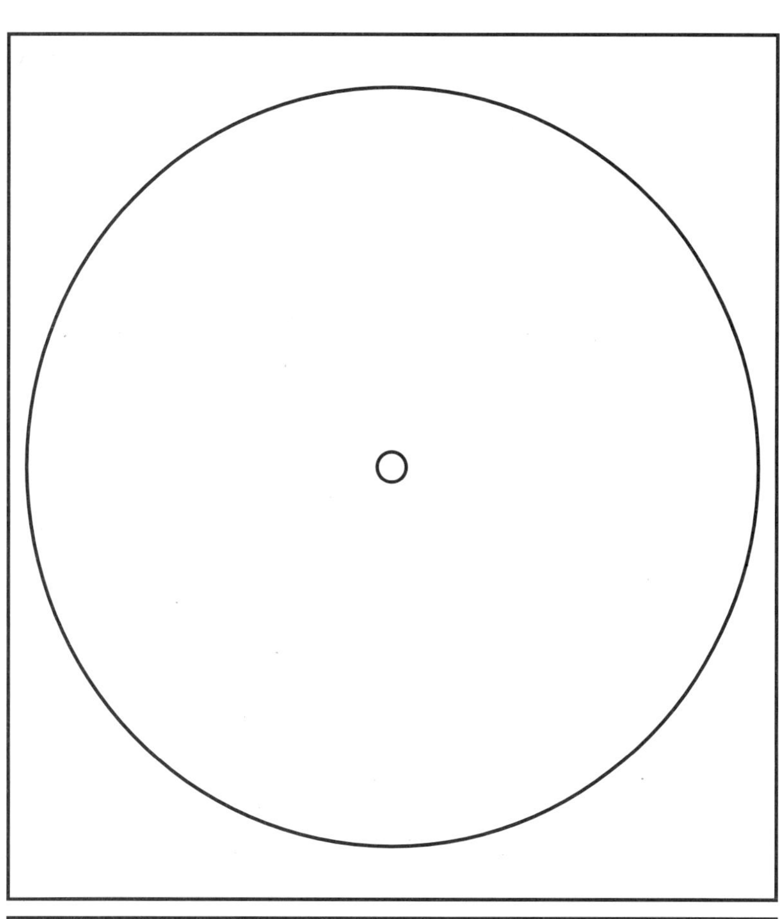

Spinner Tips

How to assemble spinner.

- Glue patterns to tagboard.
- Cut out and attach pointer with a fastener.

Alternative

- Students can use a paper clip and pencil instead.

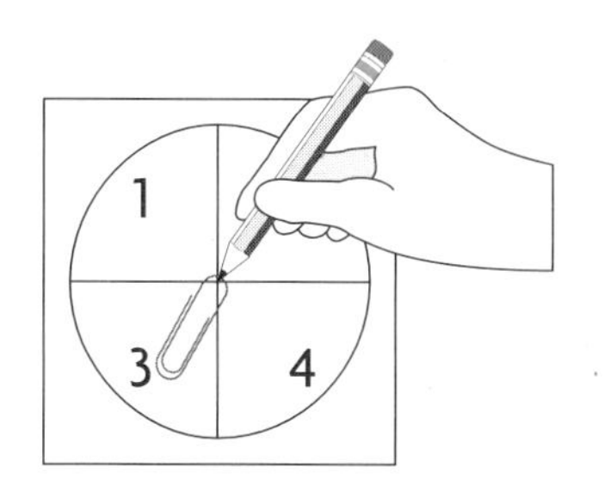

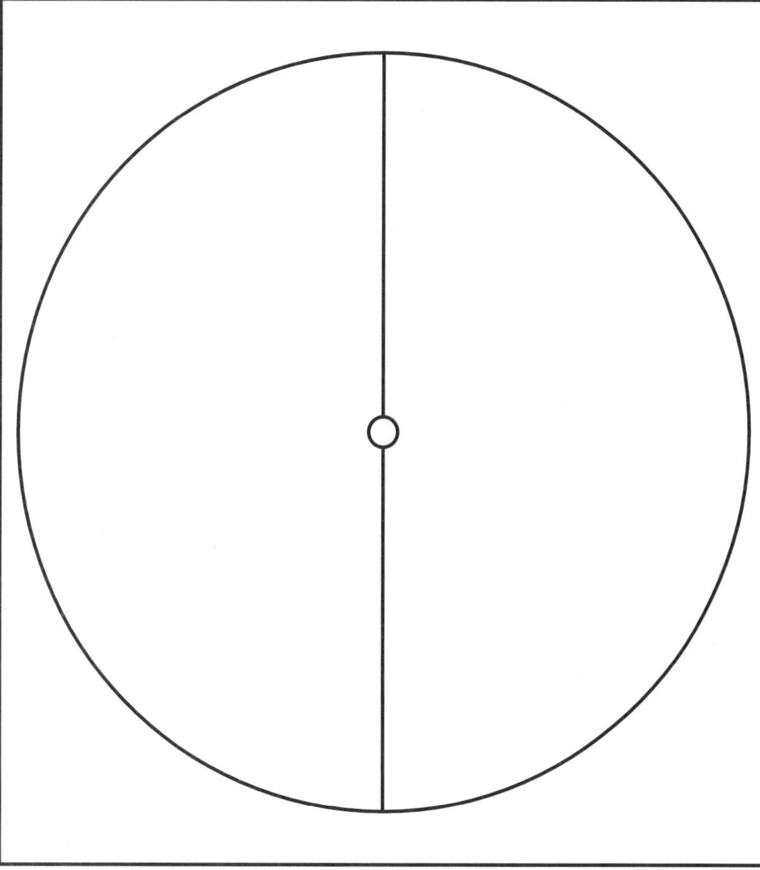

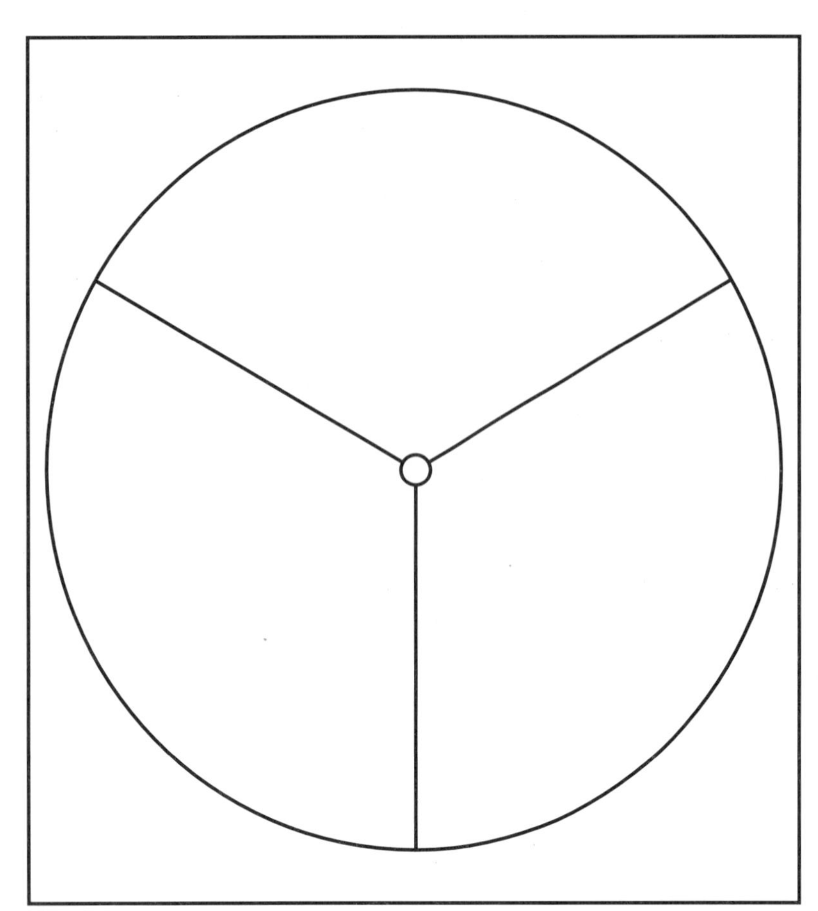

Spinner Tips

How to assemble spinner.

- Glue patterns to tagboard.
- Cut out and attach pointer with a fastener.

Alternative

- Students can use a paper clip and pencil instead.

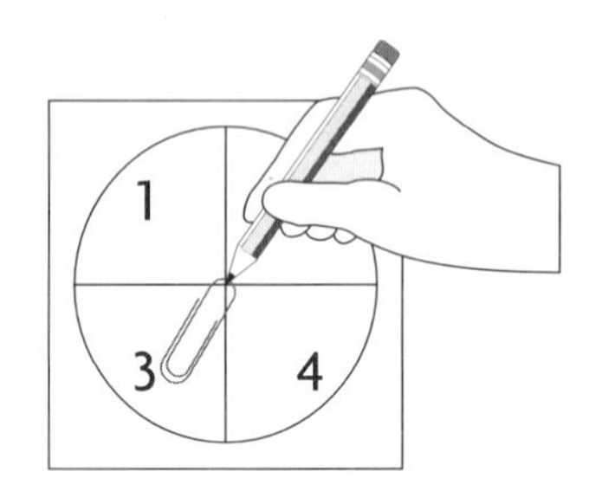

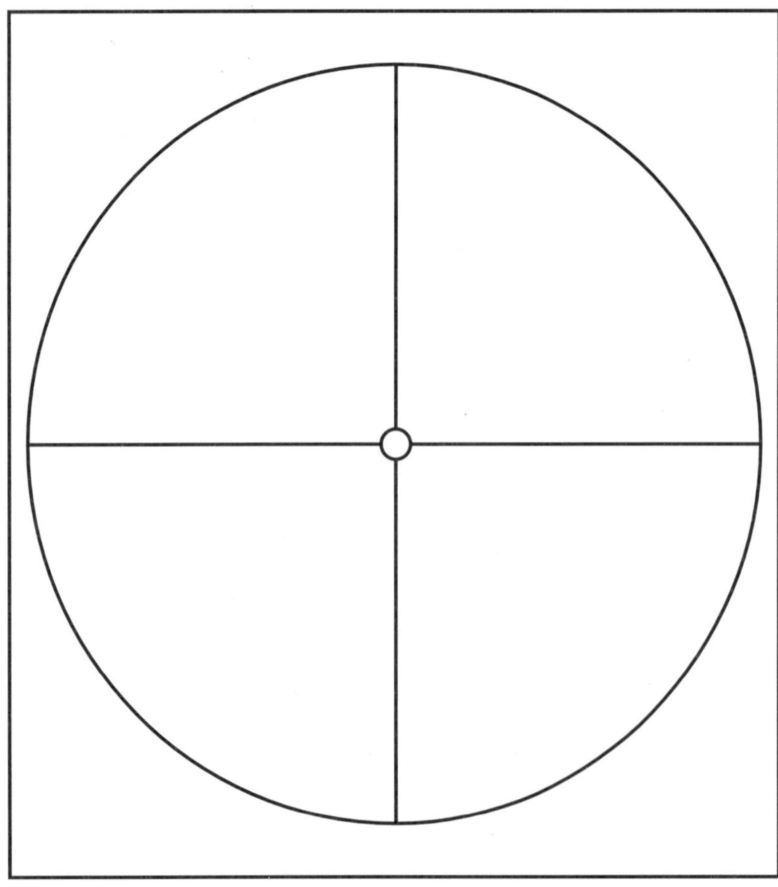

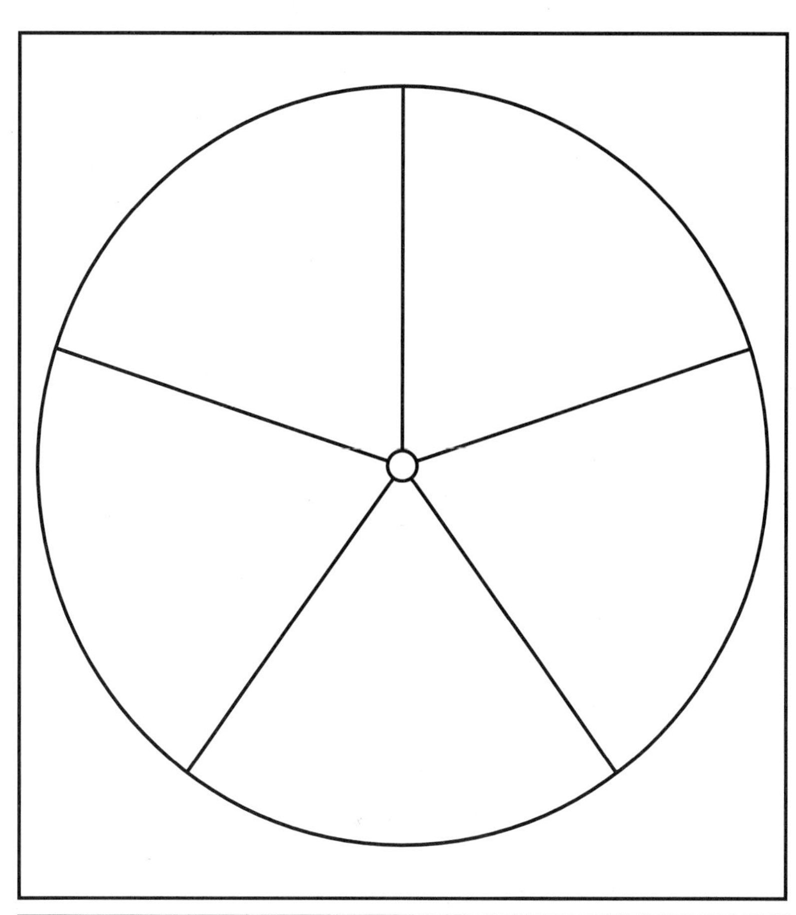

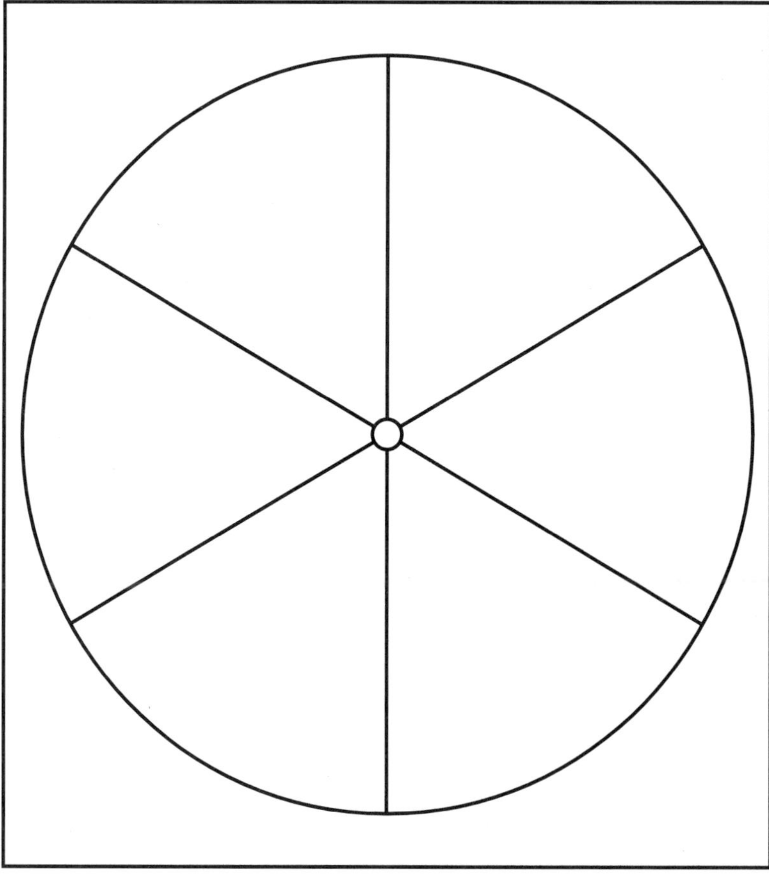

Spinner Tips

How to assemble spinner.

- Glue patterns to tagboard.
- Cut out and attach pointer with a fastener.

Alternative

- Students can use a paper clip and pencil instead.

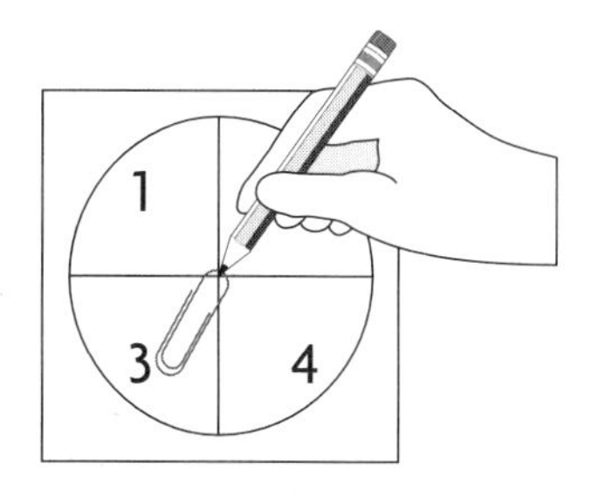

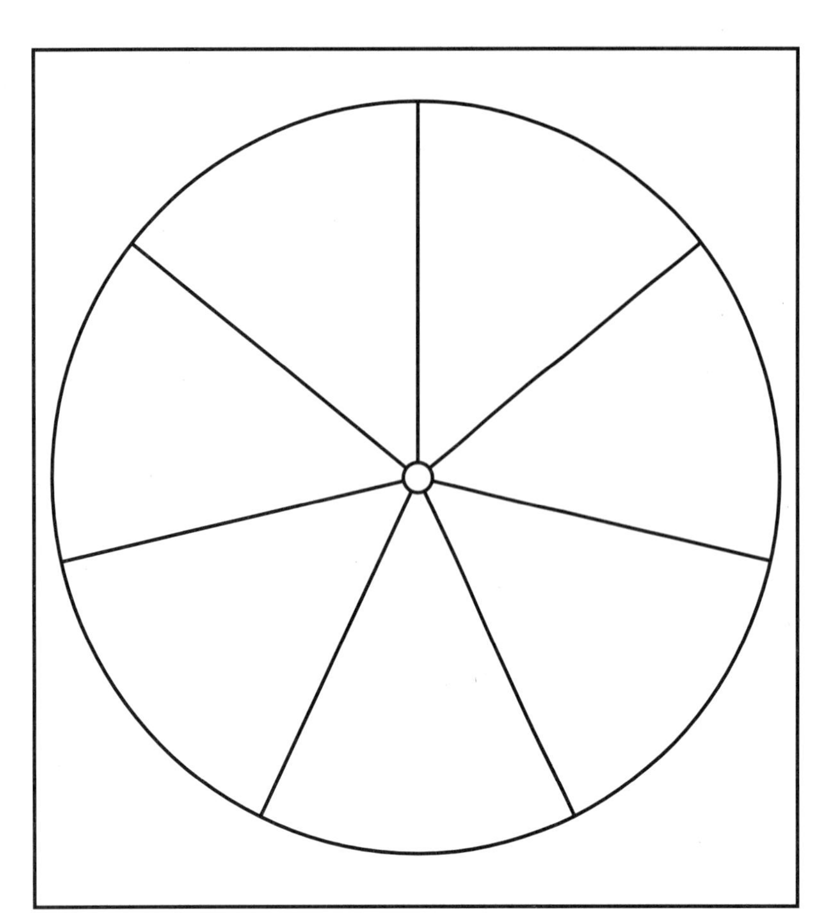

Spinner Tips

How to assemble spinner.

- Glue patterns to tagboard.
- Cut out and attach pointer with a fastener.

Alternative

- Students can use a paper clip and pencil instead.

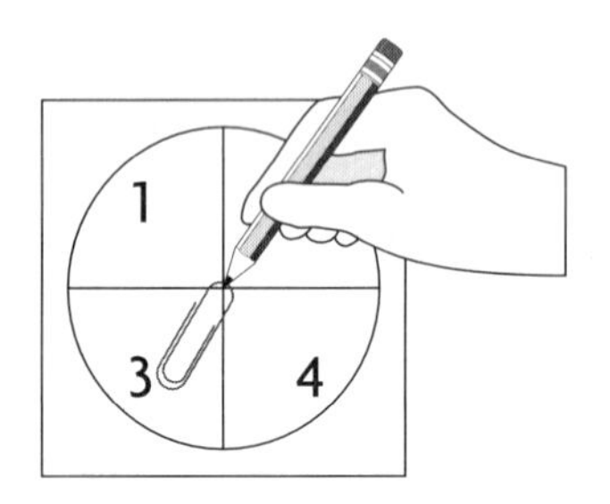

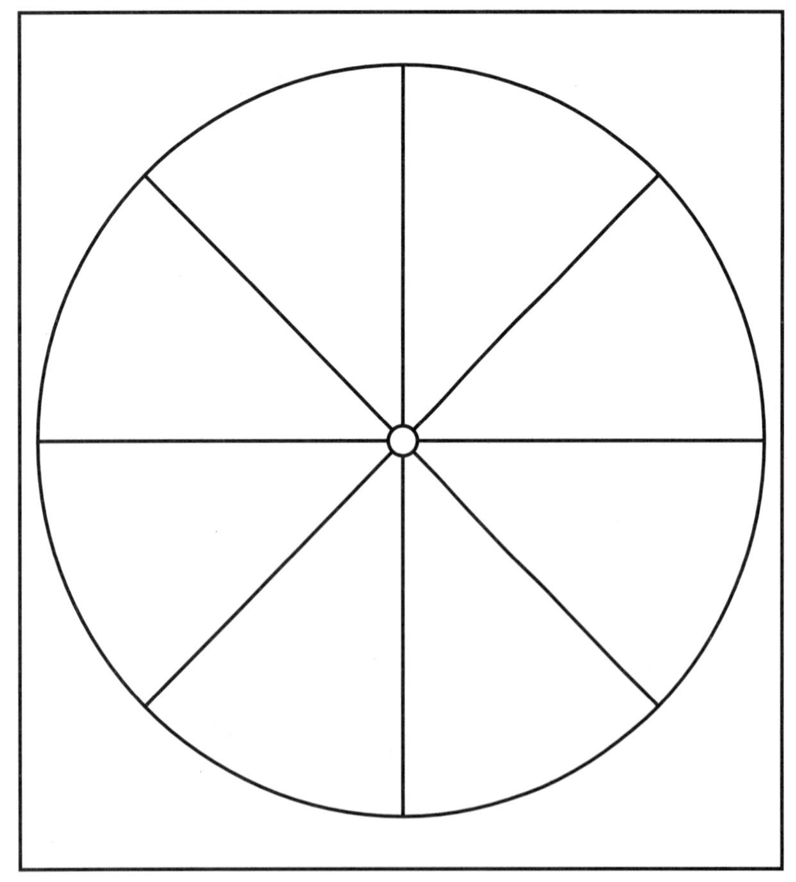

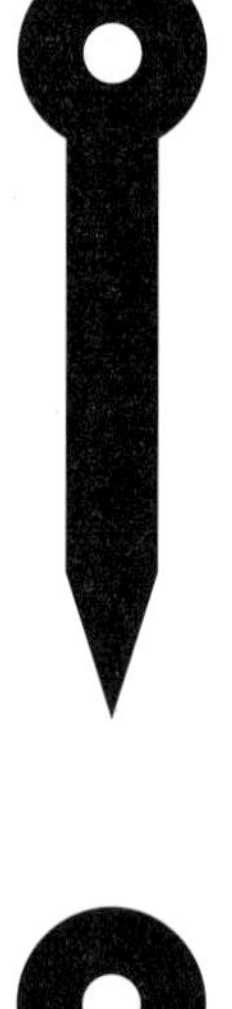

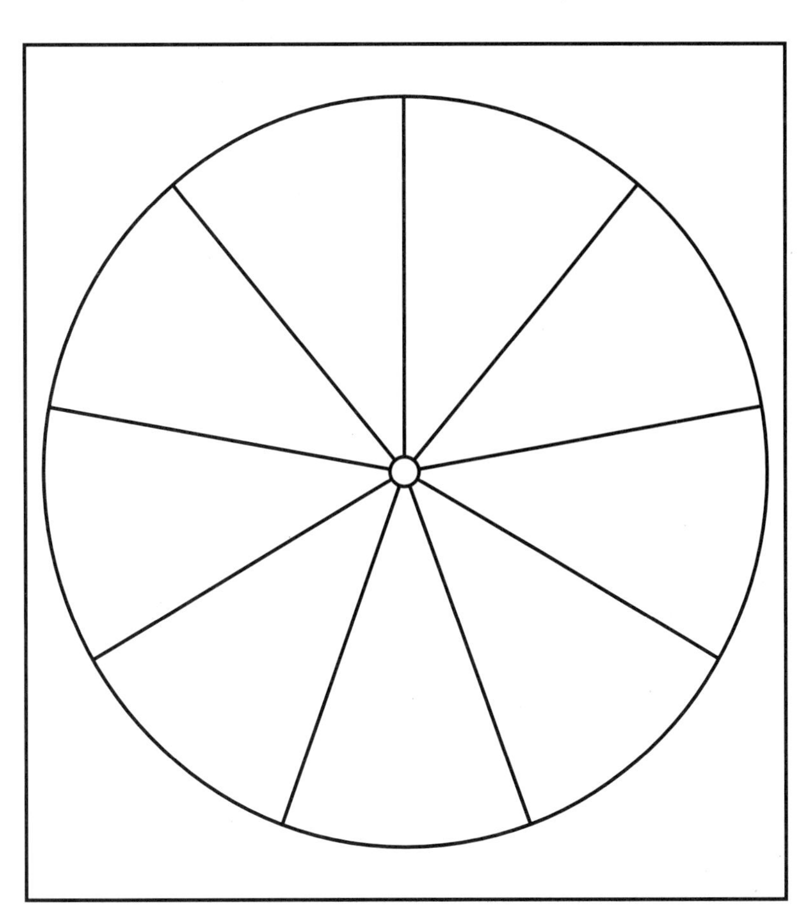

Spinner Tips

How to assemble spinner.

- Glue patterns to tagboard.
- Cut out and attach pointer with a fastener.

Alternative

- Students can use a paper clip and pencil instead.

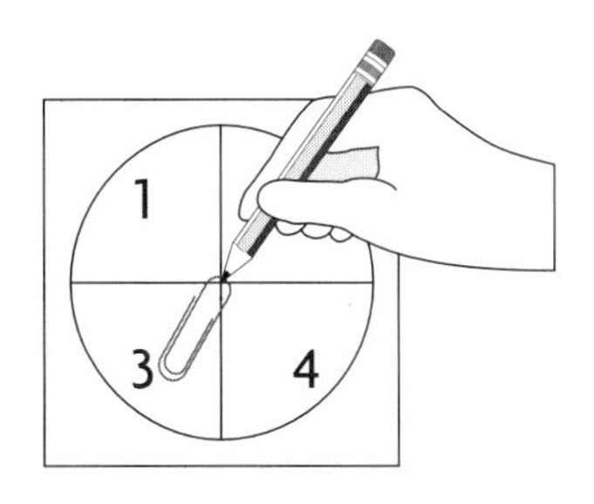

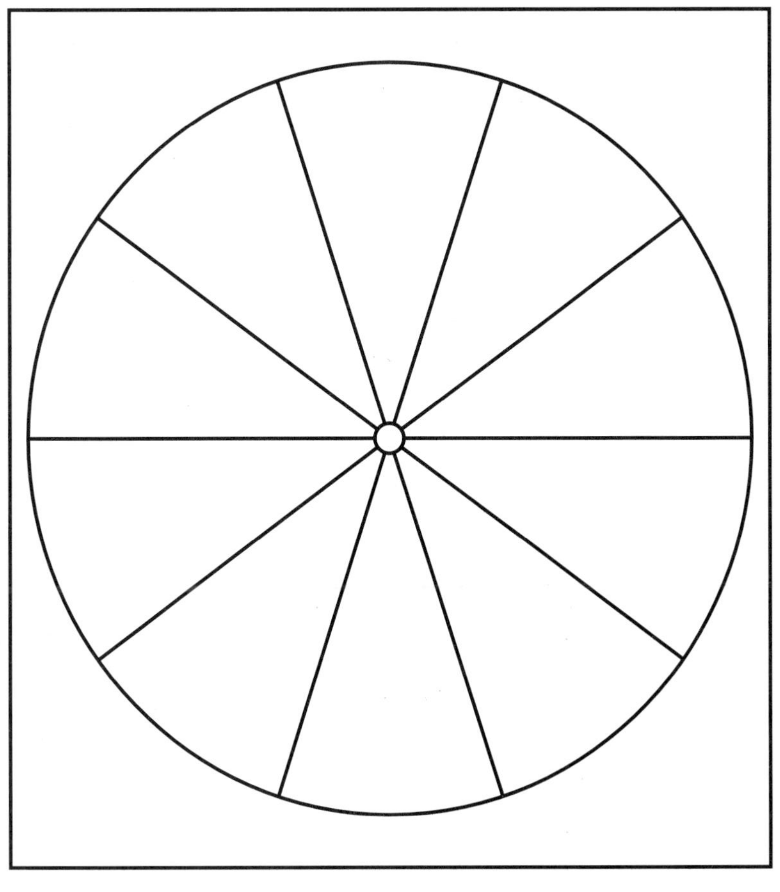

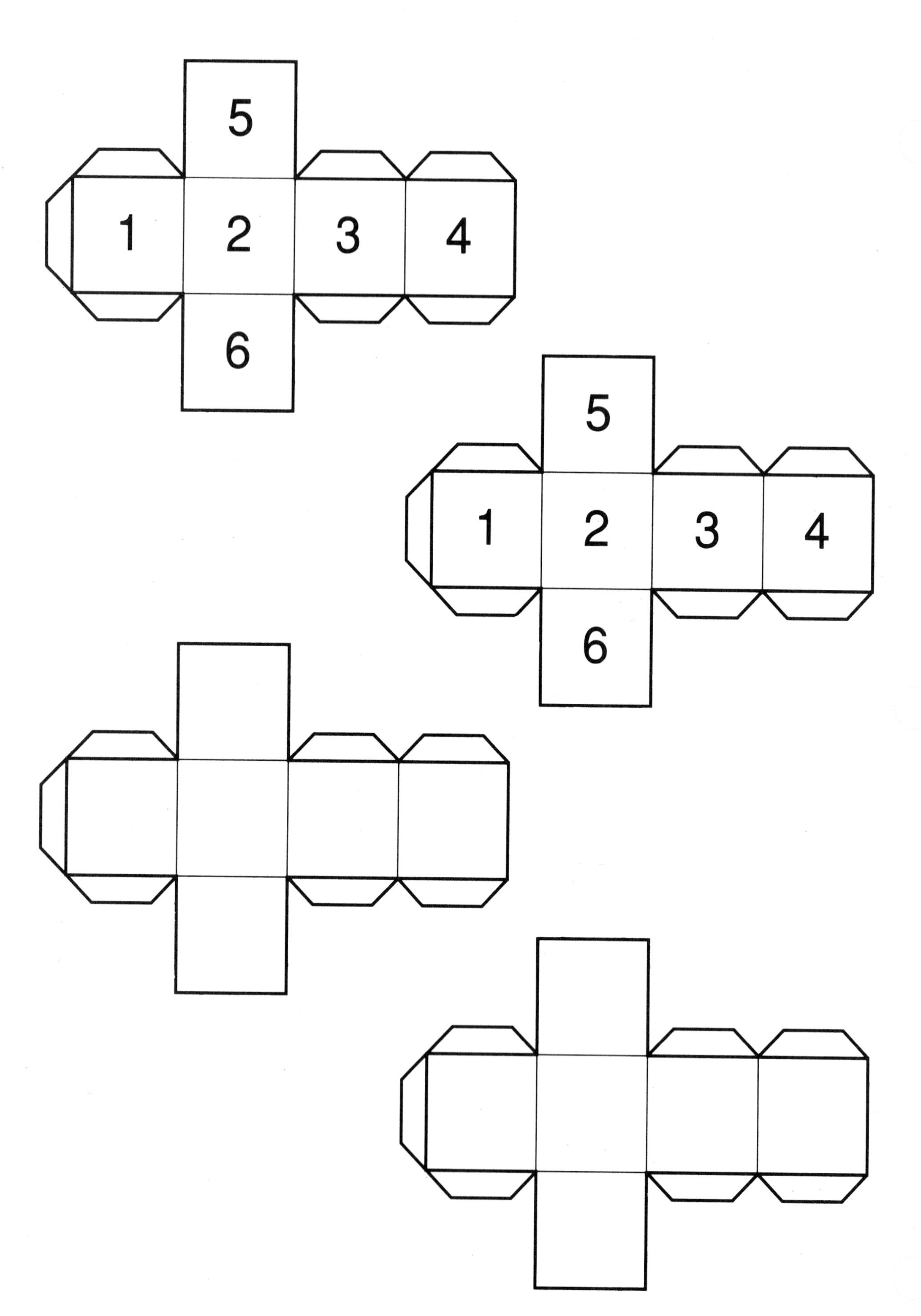
5
1
2
3
4
6
5
1
2
3
4
6

Put in order (least to greatest)

Put in order (least to greatest)

Put in order (greatest to least)

Put in order (greatest to least)

COMPARE TWO

START

901
910

862
866

Toss Again.

225
265

Go Ahead 2.

5,000
9,000

123
213

59
53

162
165

Toss Again.

330
230

549
599

Go Back 2.

2,401
2,104

9,800
8,900

600
200

67
89

Lose 1 turn.

299
266

5,432
4,321

10
101

1,991
1,091

END

Which kind of graph would best show the cost of a loaf of bread over the past 25 years? line graph	**Which kind of graph would best show the field-trip choices of fourth graders?** pictograph or bar graph	**Which kind of graph would best compare two groups, such as a group of boys and a group of girls?** double-bar graph
Which kind of graph would best show how many students were absent each month during the school year? line graph	**Which kind of graph would best show the median age of teachers in your school?** stem-and-leaf plot	**Which kind of graph would best show the number of pages each of several students could read in an hour?** bar graph
Which kind of graph would best show the means of transportation each student uses for getting to school? pictograph or bar graph	**Which kind of graph would best show the types of music enjoyed by students in a fourth-grade class?** pictograph or bar graph	**Which kind of graph would best compare the favorite sports of two classes?** double-bar graph
Which kind of graph would best show the median number of sit-ups for students in your class? stem-and-leaf plot	**Which kind of graph would best show the growth of a baby during the first year of life?** line graph	**Which kind of graph would best show the number of fourth graders who had visited famous places?** pictograph or bar graph
Which kind of graph would best show the amount of rainfall over the past year? line graph	**Which kind of graph would best show the median grade students in your class received on the math quiz?** stem-and-leaf plot	**Which kind of graph would best show the favorite authors of students in a fourth-grade class?** pictograph or bar graph

Graph Score Sheet

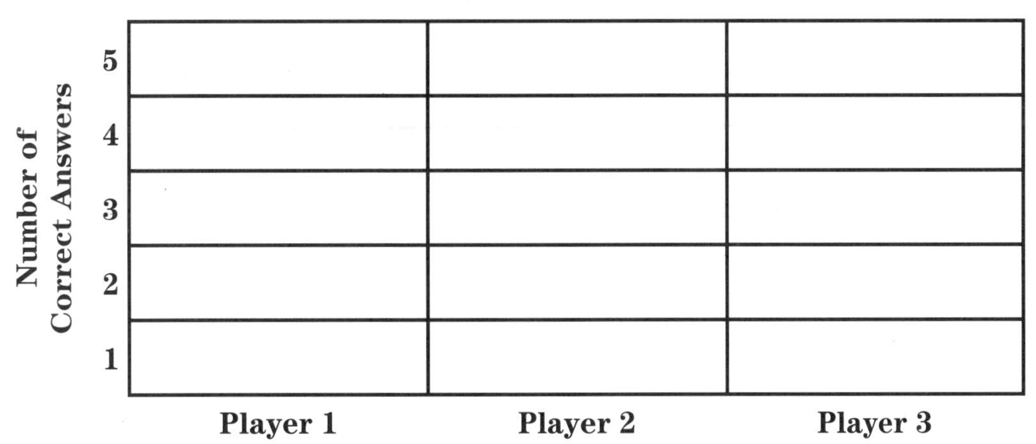

x	1	2	3	4	5	6	7	8	9
1	1	2	3	4	5	6	7	8	9
2	2	4	6	8	10	12	14	16	18
3	3	6	9	12	15	18	21	24	27
4	4	8	12	16	20	24	28	32	36
5	5	10	15	20	25	30	35	40	45
6	6	12	18	24	30	36	42	48	54
7	7	14	21	28	35	42	49	56	63
8	8	16	24	32	40	48	56	64	72
9	9	18	27	36	45	54	63	72	81

2	3
4	5
10	20
30	40
50	60

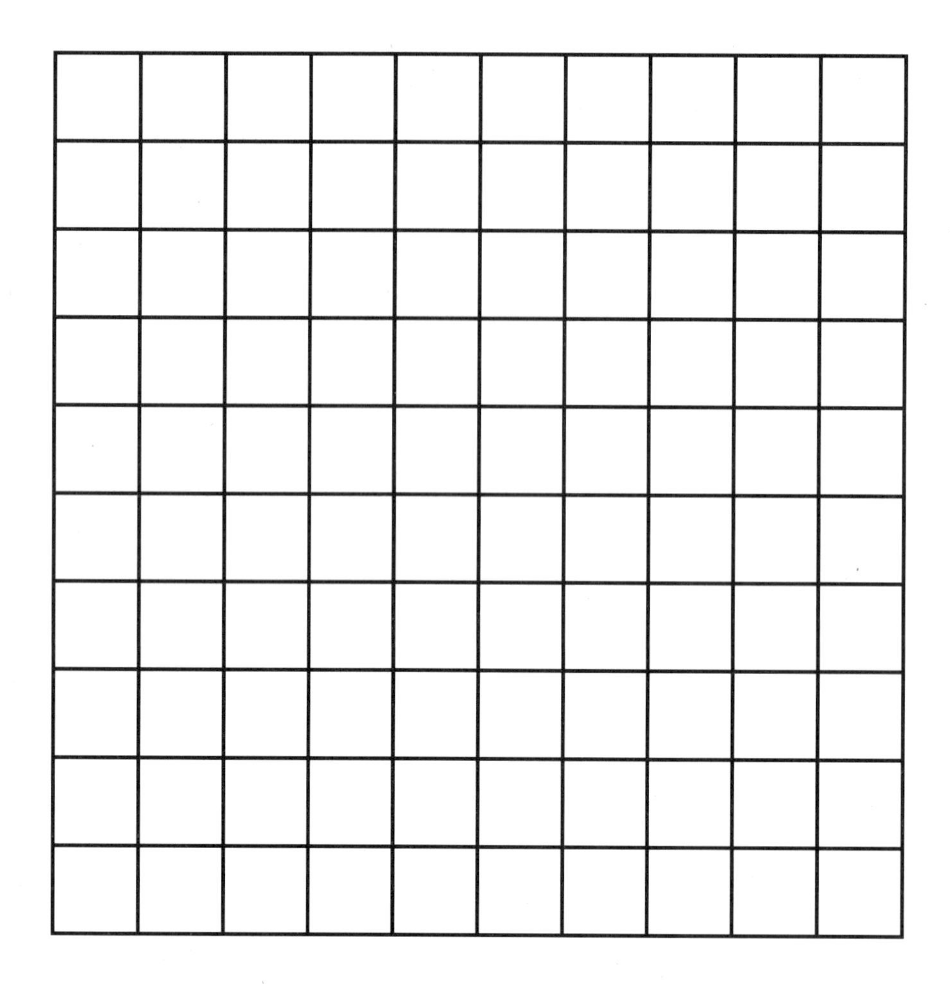

0.01	0.02	0.03	0.04	0.05
0.05	0.05	0.06	0.07	0.08
0.09	0.10	0.10	0.11	0.12
0.13	0.14	0.15	0.16	0.17
0.18	0.19	0.20	0.25	0.30

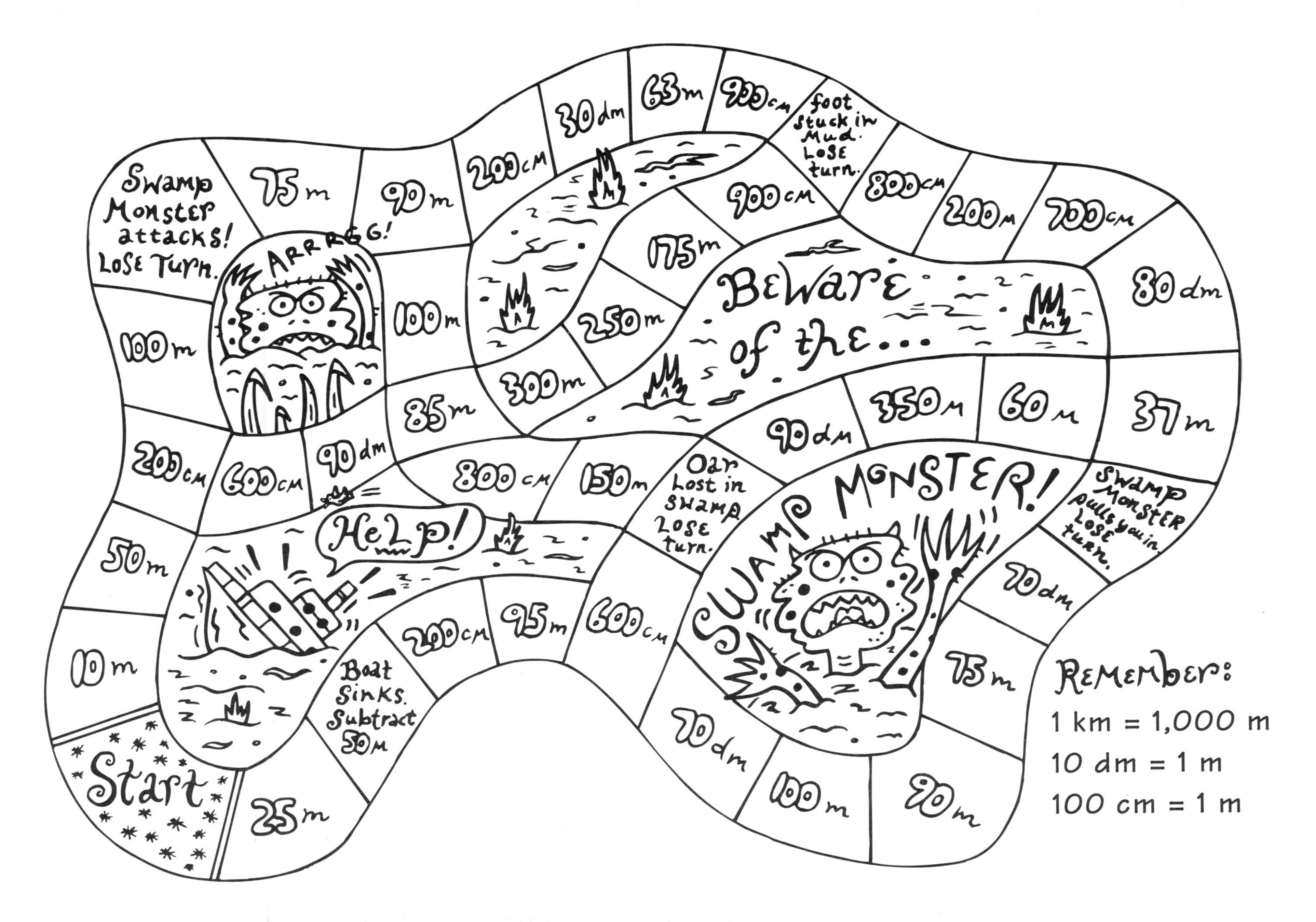
Swamp Monster attacks! Lose Turn.
75 m
90 m
200 cm
30 dm
63 m
900 cm
foot stuck in Mud. Lose turn.
800 cm
200 m
700 cm
80 dm
ARRRGG!
100 m
900 cm
175 m
Beware of the...
250 m
100 m
300 m
85 m
350 m
60 m
37 m
90 dm
200 cm
600 cm
90 dm
800 cm
150 m
Oar Lost in Swamp Lose turn.
SWAMP MONSTER!
Swamp Monster pulls you in. Lose turn.
50 m
HeLp!
70 dm
10 m
200 cm
95 m
600 cm
75 m
Boat Sinks. Subtract 50 m
Start
25 m
70 dm
100 m
90 m
Remember:
1 km = 1,000 m
10 dm = 1 m
100 cm = 1 m

Decimal Maze

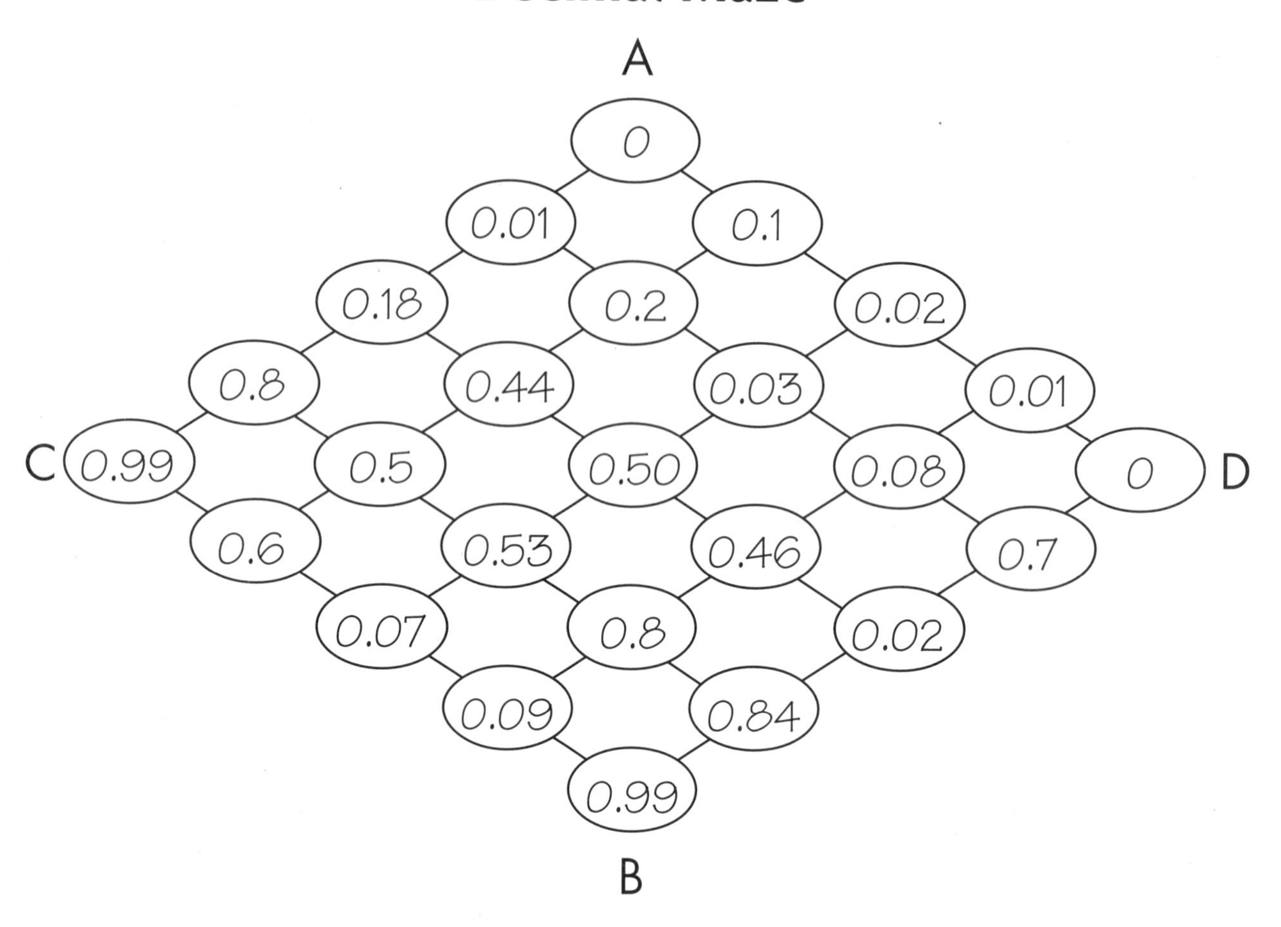

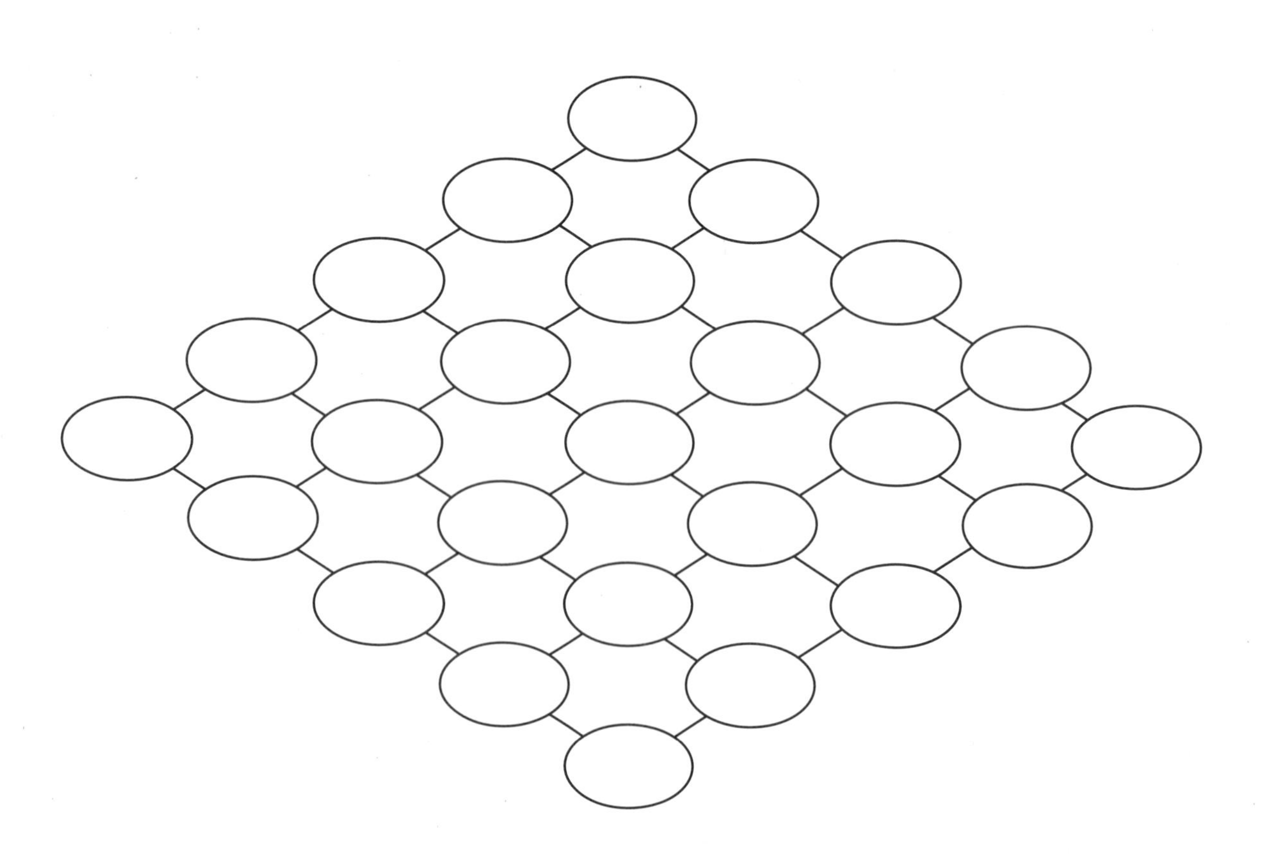

BINGO

		FREE		

Math Award

To ______________________________

For ____________________________

____

____

By ________________

EXCELLENCE

On this day of ____________

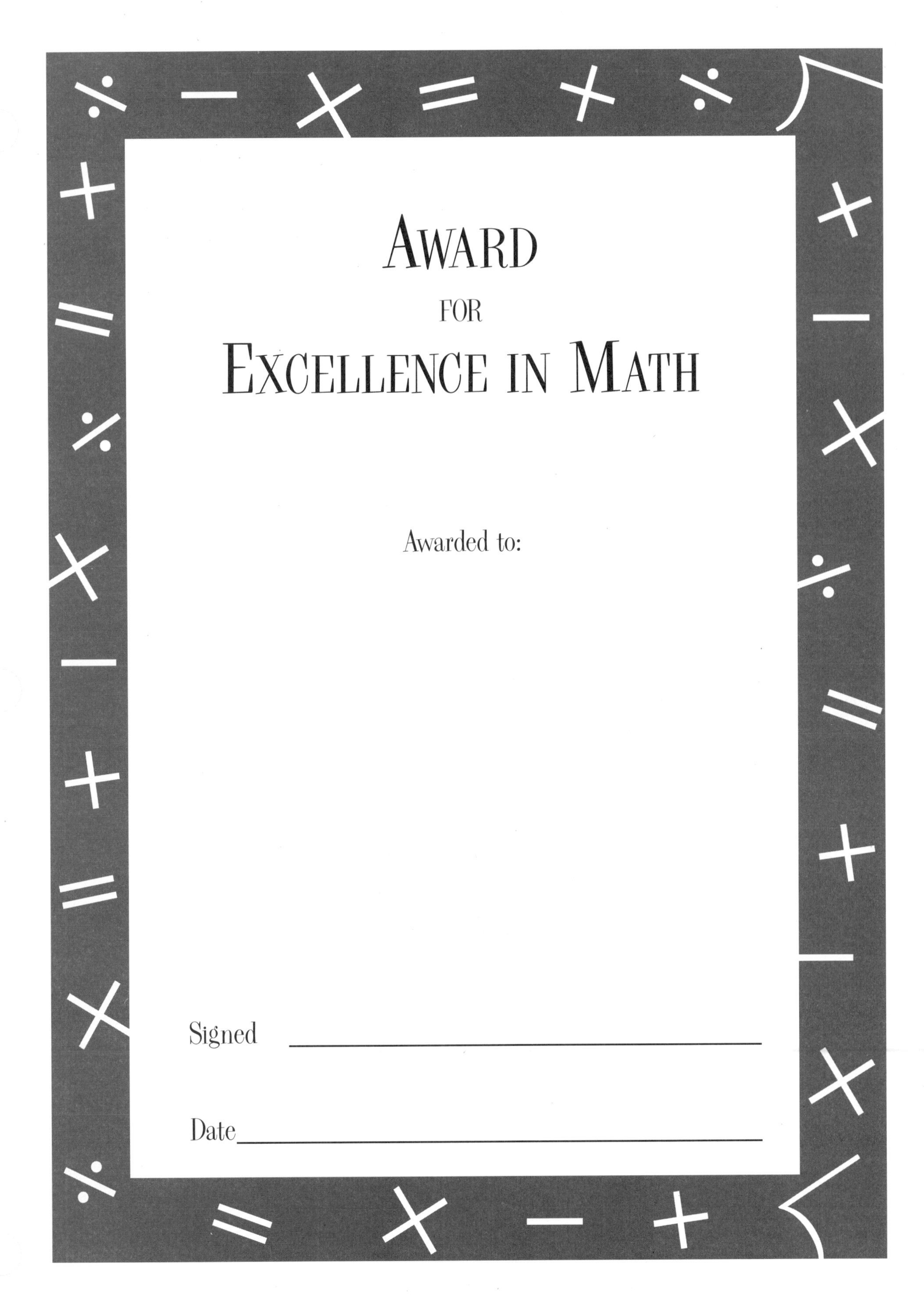
AWARD
FOR
EXCELLENCE IN MATH
Awarded to:
Signed
Date

Daily Facts Practice

A	$0 + 2$	$1 + 3$	$10 + 2$	$4 + 2$	$5 + 0$	$6 + 10$
B	$4 - 0$	$8 - 2$	$6 - 6$	$2 - 1$	$9 - 1$	$13 - 3$
C	$2 + 10$	$1 + 0$	$9 + 1$	$8 - 8$	$19 - 9$	$11 - 2$
D	1×5	9×0	7×1	3×3	0×5	2×4
E	$6\overline{)0}$	$2\overline{)2}$	$1\overline{)7}$	$1\overline{)0}$	$10\overline{)30}$	$2\overline{)10}$
F	1×8	10×4	3×2	$7\overline{)7}$	$9\overline{)0}$	$6\overline{)12}$
G	$2 + 3$	$4 + 5$	$7 + 3$	$2 + 6$	$10 + 8$	$8 + 5$

Daily Facts Practice

A	1×6	5×2	4×5	6×3	2×8	6×6
B	$10\overline{)50}$	$3\overline{)9}$	$4\overline{)28}$	0×9	8×4	10×6
C	$6 - 4$	$11 - 6$	$8 - 7$	$9 + 7$	$6 + 3$	$4 + 8$
D	$9 + 2$	$2 + 1$	$3 + 4$	$7 + 6$	$7 + 8$	$5 + 10$
E	$5 + 4$	$9 - 6$	$0 + 7$	$13 - 5$	$5 + 7$	$11 - 7$
F	$4 - 2$	$8 - 3$	$6 - 1$	$9 - 4$	$12 - 8$	$19 - 10$
G	$17 - 7$	$6 + 9$	$8 - 5$	$10 + 4$	$13 - 6$	$7 + 4$

Daily Facts Practice

A	$2 + 2$	$6 + 6$	$5 + 2$	$10 + 0$	$5 + 9$	$9 + 9$
B	0×10	2×7	5×9	6×4	6×8	3×9
C	6×5	2×3	7×0	5×5	3×8	8×9
D	$10\overline{)90}$	$6\overline{)48}$	$5\overline{)20}$	$3\overline{)18}$	$8\overline{)64}$	$7\overline{)28}$
E	$2\overline{)6}$	$3\overline{)21}$	$2\overline{)20}$	$6\overline{)36}$	$7\overline{)42}$	$9\overline{)27}$
F	$8\overline{)40}$	$3\overline{)0}$	$6\overline{)60}$	$7\overline{)14}$	$9\overline{)54}$	$8\overline{)32}$
G	$2 + 7$	$3 - 2$	$8 - 4$	$12 - 2$	$1 + 8$	$8 + 10$

Daily Facts Practice

A	$12 - 5$	$6 + 8$	$10 - 7$	$2 + 5$	$1 + 6$	$14 - 7$
B	4×6	$4\overline{)36}$	5×7	$5\overline{)30}$	9×7	$1\overline{)10}$
C	$5 + 1$	$3 + 3$	$5 + 6$	$7 + 7$	$2 + 9$	$8 + 4$
D	$3 + 9$	$10 + 10$	$0 + 6$	$3 + 8$	$6 + 5$	$4 + 7$
E	$12 - 6$	$10 - 0$	$16 - 7$	$11 - 3$	$10 - 8$	$13 - 4$
F	$9 + 3$	$10 - 10$	$3 + 10$	$14 - 9$	$8 + 8$	$12 - 7$
G	$16 - 10$	$7 - 6$	$5 - 3$	$11 - 6$	$15 - 8$	$12 - 9$

Daily Facts Practice

A	10×10	2×6	8×5	9×2	7×6	9×3
B	$7 - 4$	$16 - 8$	$11 - 1$	$15 - 9$	$12 - 10$	$18 - 9$
C	$3\overline{)30}$	8×7	$7\overline{)63}$	7×10	$9\overline{)81}$	4×4
D	$9 + 4$	$5 + 5$	$1 + 7$	$3 + 6$	$6 + 2$	$8 + 7$
E	$10 + 7$	$4 + 4$	$8 + 3$	$1 + 2$	$8 + 6$	$9 + 8$
F	$3 + 2$	$15 - 7$	$10 - 9$	$9 + 5$	$4 + 6$	$9 - 3$
G	$11 - 10$	$13 - 8$	$7 - 5$	$10 - 3$	$5 - 2$	$11 - 8$

Daily Facts Practice

A	$5 + 3$	$7 + 1$	$13 - 7$	$0 + 5$	$6 - 5$	$14 - 8$
B	$20 - 10$	$17 - 9$	$10 + 5$	$6 + 7$	$8 - 6$	$7 - 3$
C	$4 - 4$	$10 - 6$	$14 - 5$	$9 - 8$	$13 - 9$	$10 - 4$
D	7×8	4×9	4×7	$8\overline{)16}$	$5\overline{)45}$	$4\overline{)24}$
E	$4\overline{)8}$	$3\overline{)24}$	7×2	5×10	$6\overline{)24}$	$9\overline{)63}$
F	3×1	7×5	$5\overline{)50}$	$4\overline{)20}$	6×9	9×4
G	10×1	8×3	6×7	4×8	2×9	7×9

Daily Facts Practice

A	2×2	8×6	3×5	7×4	7×3	9×6
B	$2\overline{)18}$	$4\overline{)40}$	10×5	7×7	$6\overline{)30}$	$9\overline{)72}$
C	$7 + 9$	$5 + 18$	$4 + 3$	$11 - 4$	$14 - 6$	$12 - 3$
D	$2 + 4$	$4 + 9$	$10 + 3$	$2 + 9$	$7 + 5$	$8 + 2$
E	4×3	5×0	3×6	8×8	5×4	9×5
F	4×2	5×8	3×7	10×8	8×2	9×8
G	$15 - 10$	$7 - 2$	$11 - 9$	$10 - 5$	$12 - 4$	$9 - 2$

Daily Facts Practice

A	$16 - 6$	$6 - 3$	$9 - 5$	$17 - 10$	$16 - 9$	$10 - 2$
B	5×3	9×9	6×2	$2\overline{)16}$	$5\overline{)35}$	$7\overline{)49}$
C	$5\overline{)25}$	$8\overline{)72}$	$9\overline{)18}$	5×3	9×9	6×2
D	$2\overline{)8}$	2×11	$3\overline{)27}$	5×6	$5\overline{)55}$	3×4
E	2×5	11×0	12×2	10×3	7×6	6×11
F	2×6	9×1	12×4	3×7	6×9	3×8
G	8×10	11×2	4×4	8×5	6×4	5×12

Daily Facts Practice

A	0×7	$3\overline{)6}$	9×8	$9\overline{)45}$	7×3	$11\overline{)55}$
B	4×2	12×3	4×5	8×7	2×9	8×8
C	$7 + 2$	$4 - 3$	$6 + 4$	$14 - 10$	$8 + 9$	$15 - 6$
D	4×12	8×11	4×1	3×3	1×10	6×12
E	$3\overline{)36}$	$11\overline{)77}$	11×9	$7\overline{)84}$	$11\overline{)110}$	11×10
F	$4\overline{)16}$	$7\overline{)21}$	12×8	$6\overline{)54}$	$9\overline{)90}$	4×4
G	$2 + 8$	$9 + 6$	$3 + 7$	$9 - 7$	$6 - 2$	$17 - 8$

Daily Facts Practice

A	11×3	7×7	2×10	2×12	6×5	12×10
B	1×4	7×11	8×9	10×11	12×6	11×12
C	9×7	4×11	3×4	4×8	12×5	9×12
D	$8\overline{)88}$	8×2	3×12	6×6	9×3	$2\overline{)12}$
E	7×12	$8\overline{)56}$	0×8	$4\overline{)32}$	2×5	$5\overline{)60}$
F	$2\overline{)14}$	$11\overline{)44}$	$12\overline{)48}$	$6\overline{)42}$	$7\overline{)77}$	$8\overline{)48}$
G	$9\overline{)99}$	$5\overline{)0}$	$4\overline{)48}$	$7\overline{)56}$	$2\overline{)24}$	$8\overline{)80}$

Daily Facts Practice

A	$8\overline{)24}$	$2\overline{)4}$	$6\overline{)72}$	$11\overline{)121}$	$9\overline{)36}$	$10\overline{)70}$
B	$5\overline{)40}$	5×11	$6\overline{)66}$	12×9	$12\overline{)84}$	9×10
C	$3\overline{)12}$	$6\overline{)18}$	$7\overline{)35}$	$12\overline{)96}$	$11\overline{)22}$	$10\overline{)120}$
D	$18 - 10$	$5 - 4$	$11 - 6$	$15 - 7$	$10 - 4$	$7 - 1$
E	$9 - 7$	$14 - 9$	$7 - 0$	$9 - 3$	$15 - 8$	$14 - 5$
F	$3\overline{)15}$	$4\overline{)44}$	$12\overline{)72}$	$8\overline{)96}$	$7\overline{)70}$	$5\overline{)15}$
G	$5\overline{)10}$	$11\overline{)99}$	$10\overline{)100}$	$4\overline{)12}$	$7\overline{)63}$	$9\overline{)27}$

Daily Facts Practice

A	$6\overline{)30}$	$12\overline{)144}$	$3\overline{)33}$	$10\overline{)110}$	$12\overline{)36}$	$9\overline{)54}$
B	8×12	11×8	4×9	8×3	11×11	4×10
C	$13 - 10$	$7 - 4$	$11 - 2$	$15 - 5$	$11 - 3$	$14 - 6$
D	9×11	12×7	$12\overline{)24}$	9×6	10×2	$9\overline{)108}$
E	$2\overline{)22}$	$9\overline{)9}$	$12\overline{)132}$	$11\overline{)0}$	$7\overline{)28}$	$5\overline{)45}$
F	$11\overline{)66}$	$9\overline{)72}$	$4\overline{)32}$	$3\overline{)21}$	$11\overline{)11}$	$8\overline{)64}$
G	$10\overline{)40}$	$8\overline{)48}$	$6\overline{)24}$	$5\overline{)35}$	$12\overline{)120}$	$3\overline{)27}$

Daily Facts Practice

A	$12\overline{)60}$	$11\overline{)88}$	7×4	$5\overline{)25}$	$4\overline{)36}$	12×12
B	3×10	5×7	0×6	$8\overline{)8}$	$7\overline{)56}$	$2\overline{)14}$
C	10)[illegible]0	$12\overline{)108}$	$3\overline{)15}$	$6\overline{)54}$	$10\overline{)60}$	$8\overline{)96}$
D	$6 + 9$	$8 + 8$	$7 + 2$	$10 + 9$	$3 + 5$	$6 + 7$
E	$10 + 6$	$5 + 6$	$8 + 4$	$9 + 7$	$4 + 5$	$9 + 5$
F	$18 - 8$	$13 - 9$	$15 - 6$	$6 - 4$	$10 - 2$	$13 - 8$
G	$17 - 10$	$11 - 6$	$8 - 2$	$12 - 9$	$16 - 8$	$14 - 7$

Daily Facts Practice

A	3 +9	8 +6	7 +5	9 +2	4 +7	9 +9
B	12 ×11	3 ×6	2 ×4	11 ×6	3 ×9	5 ×8
C	$11\overline{)132}$	$8\overline{)24}$	3 ×11	10 ×7	$4\overline{)20}$	$9\overline{)90}$
D	$12\overline{)12}$	$8\overline{)16}$	$6\overline{)48}$	$5\overline{)15}$	$3\overline{)24}$	$12\overline{)48}$
E	$11\overline{)66}$	2 ×7	$7\overline{)21}$	10 ×12	$4\overline{)28}$	7 ×9
F	3 +8	8 +9	7 +10	7 +7	5 +9	2 +4
G	12 −5	18 −9	10 −3	14 −4	15 −9	12 −7

Daily Facts Practice

A	7 +8	8 +5	9 +10	17 −8	10 −5	13 −7
B	16 −10	13 −5	8 −3	9 +8	7 +3	0 +10
C	10 +6	8 +7	4 +6	2 +8	4 +3	7 +9
D	16 −7	9 −5	12 −4	6 −5	13 −6	11 −7
E	11 −4	9 −6	16 −9	7 +4	3 +6	9 +7
F	7 ×8	7 ×12	5 ×6	9 ×4	12 ×3	2 ×8
G	11 ×7	9 ×9	6 ×7	7 ×5	6 ×3	4 ×12

Daily Facts Practice

A	$9\overline{)63}$	$12\overline{)96}$	$5\overline{)30}$	$2\overline{)16}$	$7\overline{)42}$	$8\overline{)88}$
B	4×7	5×12	6×8	$11\overline{)33}$	$9\overline{)36}$	$6\overline{)36}$
C	$6 + 8$	$14 - 10$	$10 - 7$	$13 - 4$	$9 - 4$	$10 + 1$
D	$2 + 5$	$6 + 6$	$9 + 4$	$5 + 7$	$6 + 4$	$9 + 3$
E	$4 + 10$	$11 - 9$	$8 - 7$	$12 - 6$	$7 - 3$	$5 + 8$
F	$9 + 6$	$8 + 3$	$4 + 8$	$5 + 5$	$7 + 6$	$2 + 3$
G	$17 - 9$	$11 - 8$	$9 - 2$	$10 - 9$	$7 - 2$	$14 - 8$

Daily Facts Practice

A	$3 + 7$	$8 + 2$	$12 - 8$	$1 + 9$	$10 - 6$	$7 - 5$
B	$4 + 9$	$8 - 5$	$10 - 8$	$6 + 5$	$4 + 2$	$12 - 3$
C	12×4	5×9	8×4	4×3	11×5	8×6
D	6×2	12×9	3×5	9×2	8×8	11×4
E	$12\overline{)120}$	$9\overline{)18}$	$6\overline{)72}$	$3\overline{)18}$	$8\overline{)80}$	$6\overline{)60}$
F	$2\overline{)18}$	$5\overline{)60}$	$6\overline{)18}$	$8\overline{)56}$	$10\overline{)20}$	$12\overline{)84}$
G	$9\overline{)45}$	$7\overline{)49}$	$12\overline{)108}$	$7\overline{)14}$	$4\overline{)24}$	$6\overline{)42}$

Daily Facts Practice

A	4×6	$12\overline{)72}$	9×5	$10\overline{)0}$	$9\overline{)81}$	$4\overline{)48}$
B	$2 + 9$	$5 + 4$	$6 + 2$	$8 + 6$	$7 + 5$	$7 + 9$
C	$8 - 6$	$2 + 6$	$6 - 2$	$14 - 5$	$8 + 7$	$10 - 7$
D	$5 + 2$	$7 + 4$	$5 + 8$	$3 + 7$	$6 + 3$	$10 + 9$
E	$15 - 10$	$11 - 8$	$5 - 3$	$8 - 5$	$14 - 9$	$9 - 3$
F	$3 + 5$	$13 - 7$	$6 - 3$	$9 + 5$	$10 - 4$	$14 - 6$
G	6×0	5×2	5×5	7×6	7×8	8×12

Daily Facts Practice

A	7×11	12×12	4×9	5×7	8×6	5×4
B	$7\overline{)35}$	$8\overline{)32}$	$9\overline{)108}$	$10\overline{)10}$	$11\overline{)121}$	$12\overline{)144}$
C	$8\overline{)40}$	$8\overline{)72}$	$4\overline{)16}$	$3\overline{)36}$	$12\overline{)132}$	$5\overline{)40}$
D	6×9	4×7	12×5	$12\overline{)60}$	$5\overline{)20}$	$7\overline{)28}$
E	$2 + 7$	$4 + 4$	$6 + 10$	$8 + 9$	$3 + 8$	$5 + 6$
F	$8 - 4$	$9 - 5$	$14 - 7$	$17 - 8$	$9 - 2$	$18 - 10$
G	$5 + 9$	$7 + 7$	$10 + 4$	$16 - 8$	$12 - 6$	$6 - 3$

Daily Facts Practice

A	9×12	6×5	3×7	11×11	4×3	4×8
B	$8\overline{)80}$	$6\overline{)12}$	$12\overline{)36}$	$9\overline{)63}$	$4\overline{)44}$	$8\overline{)72}$
C	6×4	$7\overline{)84}$	9×3	$3\overline{)12}$	6×7	$7\overline{)49}$
D	3×9	5×8	4×6	11×12	8×4	9×5
E	5×11	10×9	7×7	12×15	6×6	5×3
F	$8\overline{)64}$	$5\overline{)55}$	$6\overline{)30}$	$4\overline{)32}$	$2\overline{)24}$	$6\overline{)54}$
G	$5\overline{)25}$	3×12	$7\overline{)21}$	8×2	$12\overline{)96}$	8×9

Daily Facts Practice

A	$8 + 8$	$4 + 10$	$7 + 6$	$4 + 5$	$3 + 4$	$9 + 3$
B	$17 - 9$	$12 - 5$	$7 - 7$	$13 - 4$	$8 - 2$	$13 - 8$
C	$3 + 9$	$6 + 6$	$2 + 5$	$7 + 3$	$6 + 7$	$4 + 8$
D	$11 - 9$	$7 - 6$	$10 - 3$	$10 - 6$	$15 - 8$	$8 - 4$
E	$6\overline{)24}$	$9\overline{)27}$	$7\overline{)77}$	$6\overline{)42}$	$4\overline{)36}$	$2\overline{)18}$
F	$12\overline{)24}$	$8\overline{)32}$	$3\overline{)27}$	$2\overline{)20}$	$8\overline{)24}$	$5\overline{)45}$
G	4×4	8×3	$2\overline{)12}$	$4\overline{)28}$	6×12	$9\overline{)99}$

Daily Facts Practice

A	12×7	5×9	7×4	5×5	8×11	6×3
B	10×0	3×4	12×11	4×5	5×6	9×6
C	$5\overline{)20}$	3×8	$9\overline{)81}$	$12\overline{)0}$	2×12	$8\overline{)56}$
D	$3 + 2$	$9 + 4$	$6 + 5$	$5 + 3$	$8 + 4$	$6 + 1$
E	$16 - 9$	$11 - 7$	$15 - 6$	$9 - 8$	$10 - 2$	$13 - 9$
F	$9 - 4$	$8 - 6$	$5 + 7$	$16 - 7$	$9 + 9$	$13 - 5$
G	7×12	6×2	2×9	8×5	12×8	7×9

Daily Facts Practice

A	6×8	2×7	3×5	9×4	12×6	9×8
B	$7\overline{)14}$	$9\overline{)54}$	$6\overline{)66}$	$5\overline{)35}$	$3\overline{)24}$	$7\overline{)63}$
C	$2\overline{)8}$	$4\overline{)24}$	$9\overline{)36}$	$10\overline{)110}$	$6\overline{)72}$	$7\overline{)0}$
D	$9\overline{)45}$	$8\overline{)16}$	$4\overline{)20}$	$12\overline{)72}$	$3\overline{)36}$	$4\overline{)40}$
E	7×5	$8\overline{)40}$	12×1	$5\overline{)30}$	8×7	$7\overline{)70}$
F	$9 + 8$	$6 + 9$	$8 + 5$	$7 + 8$	$4 + 9$	$4 + 7$
G	$18 - 9$	$13 - 6$	$12 - 4$	$11 - 2$	$12 - 9$	$14 - 8$

Daily Facts Practice

A	$\begin{array}{r} 6 \\ +8 \\ \hline \end{array}$	$\begin{array}{r} 4 \\ +6 \\ \hline \end{array}$	$\begin{array}{r} 8 \\ +3 \\ \hline \end{array}$	$\begin{array}{r} 15 \\ -7 \\ \hline \end{array}$	$\begin{array}{r} 11 \\ -4 \\ \hline \end{array}$	$\begin{array}{r} 12 \\ -7 \\ \hline \end{array}$
B	$\begin{array}{r} 2 \\ \times 8 \\ \hline \end{array}$	$\begin{array}{r} 6 \\ \times 12 \\ \hline \end{array}$	$\begin{array}{r} 9 \\ \times 9 \\ \hline \end{array}$	$\begin{array}{r} 3 \\ \times 6 \\ \hline \end{array}$	$\begin{array}{r} 9 \\ \times 7 \\ \hline \end{array}$	$\begin{array}{r} 5 \\ \times 4 \\ \hline \end{array}$
C	$\begin{array}{r} 6 \\ \times 10 \\ \hline \end{array}$	$\begin{array}{r} 3 \\ \times 12 \\ \hline \end{array}$	$\begin{array}{r} 0 \\ \times 11 \\ \hline \end{array}$	$6\overline{)48}$	$9\overline{)72}$	$7\overline{)42}$

$\begin{array}{r} 0 \\ \times 0 \\ \hline \end{array}$	$\begin{array}{r} 1 \\ \times 0 \\ \hline \end{array}$	$\begin{array}{r} 2 \\ \times 0 \\ \hline \end{array}$
$\begin{array}{r} 3 \\ \times 0 \\ \hline \end{array}$	$\begin{array}{r} 4 \\ \times 0 \\ \hline \end{array}$	$\begin{array}{r} 5 \\ \times 0 \\ \hline \end{array}$
$\begin{array}{r} 6 \\ \times 0 \\ \hline \end{array}$	$\begin{array}{r} 7 \\ \times 0 \\ \hline \end{array}$	$\begin{array}{r} 8 \\ \times 0 \\ \hline \end{array}$

$\begin{array}{r} 9 \\ \times 0 \\ \hline \end{array}$	$\begin{array}{r} 10 \\ \times 0 \\ \hline \end{array}$	$\begin{array}{r} 11 \\ \times 0 \\ \hline \end{array}$
$\begin{array}{r} 12 \\ \times 0 \\ \hline \end{array}$	$\begin{array}{r} 0 \\ \times 1 \\ \hline \end{array}$	$\begin{array}{r} 1 \\ \times 1 \\ \hline \end{array}$
$\begin{array}{r} 2 \\ \times 1 \\ \hline \end{array}$	$\begin{array}{r} 3 \\ \times 1 \\ \hline \end{array}$	$\begin{array}{r} 4 \\ \times 1 \\ \hline \end{array}$

$\begin{array}{r} 5 \\ \times 1 \\ \hline \end{array}$	$\begin{array}{r} 6 \\ \times 1 \\ \hline \end{array}$	$\begin{array}{r} 7 \\ \times 1 \\ \hline \end{array}$
$\begin{array}{r} 8 \\ \times 1 \\ \hline \end{array}$	$\begin{array}{r} 9 \\ \times 1 \\ \hline \end{array}$	$\begin{array}{r} 10 \\ \times 1 \\ \hline \end{array}$
$\begin{array}{r} 11 \\ \times 1 \\ \hline \end{array}$	$\begin{array}{r} 12 \\ \times 1 \\ \hline \end{array}$	$\begin{array}{r} 0 \\ \times 2 \\ \hline \end{array}$

$\begin{array}{r} 1 \\ \times 2 \\ \hline \end{array}$	$\begin{array}{r} 2 \\ \times 2 \\ \hline \end{array}$	$\begin{array}{r} 3 \\ \times 2 \\ \hline \end{array}$
$\begin{array}{r} 4 \\ \times 2 \\ \hline \end{array}$	$\begin{array}{r} 5 \\ \times 2 \\ \hline \end{array}$	$\begin{array}{r} 6 \\ \times 2 \\ \hline \end{array}$
$\begin{array}{r} 7 \\ \times 2 \\ \hline \end{array}$	$\begin{array}{r} 8 \\ \times 2 \\ \hline \end{array}$	$\begin{array}{r} 9 \\ \times 2 \\ \hline \end{array}$

10×2	11×2	12×2
0×3	1×3	2×3
3×3	4×3	5×3

$\begin{array}{r} 6 \\ \times 3 \\ \hline \end{array}$	$\begin{array}{r} 7 \\ \times 3 \\ \hline \end{array}$	$\begin{array}{r} 8 \\ \times 3 \\ \hline \end{array}$
$\begin{array}{r} 9 \\ \times 3 \\ \hline \end{array}$	$\begin{array}{r} 10 \\ \times 3 \\ \hline \end{array}$	$\begin{array}{r} 11 \\ \times 3 \\ \hline \end{array}$
$\begin{array}{r} 12 \\ \times 3 \\ \hline \end{array}$	$\begin{array}{r} 0 \\ \times 4 \\ \hline \end{array}$	$\begin{array}{r} 1 \\ \times 4 \\ \hline \end{array}$

$\begin{array}{r} 2 \\ \times 4 \\ \hline \end{array}$	$\begin{array}{r} 3 \\ \times 4 \\ \hline \end{array}$	$\begin{array}{r} 4 \\ \times 4 \\ \hline \end{array}$
$\begin{array}{r} 5 \\ \times 4 \\ \hline \end{array}$	$\begin{array}{r} 6 \\ \times 4 \\ \hline \end{array}$	$\begin{array}{r} 7 \\ \times 4 \\ \hline \end{array}$
$\begin{array}{r} 8 \\ \times 4 \\ \hline \end{array}$	$\begin{array}{r} 9 \\ \times 4 \\ \hline \end{array}$	$\begin{array}{r} 10 \\ \times 4 \\ \hline \end{array}$

$\begin{array}{r} 11 \\ \times 4 \\ \hline \end{array}$	$\begin{array}{r} 12 \\ \times 4 \\ \hline \end{array}$	$\begin{array}{r} 0 \\ \times 5 \\ \hline \end{array}$
$\begin{array}{r} 1 \\ \times 5 \\ \hline \end{array}$	$\begin{array}{r} 2 \\ \times 5 \\ \hline \end{array}$	$\begin{array}{r} 3 \\ \times 5 \\ \hline \end{array}$
$\begin{array}{r} 4 \\ \times 5 \\ \hline \end{array}$	$\begin{array}{r} 5 \\ \times 5 \\ \hline \end{array}$	$\begin{array}{r} 6 \\ \times 5 \\ \hline \end{array}$

$\begin{array}{r} 7 \\ \times 5 \\ \hline \end{array}$	$\begin{array}{r} 8 \\ \times 5 \\ \hline \end{array}$	$\begin{array}{r} 9 \\ \times 5 \\ \hline \end{array}$
$\begin{array}{r} 10 \\ \times 5 \\ \hline \end{array}$	$\begin{array}{r} 11 \\ \times 6 \\ \hline \end{array}$	$\begin{array}{r} 12 \\ \times 5 \\ \hline \end{array}$
$\begin{array}{r} 0 \\ \times 6 \\ \hline \end{array}$	$\begin{array}{r} 1 \\ \times 6 \\ \hline \end{array}$	$\begin{array}{r} 2 \\ \times 6 \\ \hline \end{array}$

$\begin{array}{r} 3 \\ \times 6 \\ \hline \end{array}$	$\begin{array}{r} 4 \\ \times 6 \\ \hline \end{array}$	$\begin{array}{r} 5 \\ \times 6 \\ \hline \end{array}$
$\begin{array}{r} 6 \\ \times 6 \\ \hline \end{array}$	$\begin{array}{r} 7 \\ \times 6 \\ \hline \end{array}$	$\begin{array}{r} 8 \\ \times 6 \\ \hline \end{array}$
$\begin{array}{r} 9 \\ \times 6 \\ \hline \end{array}$	$\begin{array}{r} 10 \\ \times 6 \\ \hline \end{array}$	$\begin{array}{r} 11 \\ \times 6 \\ \hline \end{array}$

$$\begin{array}{r} 12 \\ \times 6 \\ \hline \end{array}$$

$$\begin{array}{r} 0 \\ \times 7 \\ \hline \end{array}$$

$$\begin{array}{r} 1 \\ \times 7 \\ \hline \end{array}$$

$$\begin{array}{r} 2 \\ \times 7 \\ \hline \end{array}$$

$$\begin{array}{r} 3 \\ \times 7 \\ \hline \end{array}$$

$$\begin{array}{r} 4 \\ \times 7 \\ \hline \end{array}$$

$$\begin{array}{r} 5 \\ \times 7 \\ \hline \end{array}$$

$$\begin{array}{r} 6 \\ \times 7 \\ \hline \end{array}$$

$$\begin{array}{r} 7 \\ \times 7 \\ \hline \end{array}$$

$\begin{array}{r} 8 \\ \times 7 \\ \hline \end{array}$	$\begin{array}{r} 9 \\ \times 7 \\ \hline \end{array}$	$\begin{array}{r} 10 \\ \times 7 \\ \hline \end{array}$
$\begin{array}{r} 11 \\ \times 7 \\ \hline \end{array}$	$\begin{array}{r} 12 \\ \times 7 \\ \hline \end{array}$	$\begin{array}{r} 0 \\ \times 8 \\ \hline \end{array}$
$\begin{array}{r} 1 \\ \times 8 \\ \hline \end{array}$	$\begin{array}{r} 2 \\ \times 8 \\ \hline \end{array}$	$\begin{array}{r} 3 \\ \times 8 \\ \hline \end{array}$

$\begin{array}{r} 4 \\ \times 8 \\ \hline \end{array}$	$\begin{array}{r} 5 \\ \times 8 \\ \hline \end{array}$	$\begin{array}{r} 6 \\ \times 8 \\ \hline \end{array}$
$\begin{array}{r} 7 \\ \times 8 \\ \hline \end{array}$	$\begin{array}{r} 8 \\ \times 8 \\ \hline \end{array}$	$\begin{array}{r} 9 \\ \times 8 \\ \hline \end{array}$
$\begin{array}{r} 10 \\ \times 8 \\ \hline \end{array}$	$\begin{array}{r} 11 \\ \times 8 \\ \hline \end{array}$	$\begin{array}{r} 12 \\ \times 8 \\ \hline \end{array}$

$\begin{array}{r} 0 \\ \times 9 \\ \hline \end{array}$	$\begin{array}{r} 1 \\ \times 9 \\ \hline \end{array}$	$\begin{array}{r} 2 \\ \times 9 \\ \hline \end{array}$
$\begin{array}{r} 3 \\ \times 9 \\ \hline \end{array}$	$\begin{array}{r} 4 \\ \times 9 \\ \hline \end{array}$	$\begin{array}{r} 5 \\ \times 9 \\ \hline \end{array}$
$\begin{array}{r} 6 \\ \times 9 \\ \hline \end{array}$	$\begin{array}{r} 7 \\ \times 9 \\ \hline \end{array}$	$\begin{array}{r} 8 \\ \times 9 \\ \hline \end{array}$

$$\begin{array}{r} 9 \\ \times 9 \\ \hline \end{array}$$

$$\begin{array}{r} 10 \\ \times 9 \\ \hline \end{array}$$

$$\begin{array}{r} 11 \\ \times 9 \\ \hline \end{array}$$

$$\begin{array}{r} 12 \\ \times 9 \\ \hline \end{array}$$

$$\begin{array}{r} 0 \\ \times 10 \\ \hline \end{array}$$

$$\begin{array}{r} 1 \\ \times 10 \\ \hline \end{array}$$

$$\begin{array}{r} 2 \\ \times 10 \\ \hline \end{array}$$

$$\begin{array}{r} 3 \\ \times 10 \\ \hline \end{array}$$

$$\begin{array}{r} 4 \\ \times 10 \\ \hline \end{array}$$

$\begin{array}{r} 5 \\ \times 10 \\ \hline \end{array}$	$\begin{array}{r} 6 \\ \times 10 \\ \hline \end{array}$	$\begin{array}{r} 7 \\ \times 10 \\ \hline \end{array}$
$\begin{array}{r} 8 \\ \times 10 \\ \hline \end{array}$	$\begin{array}{r} 9 \\ \times 10 \\ \hline \end{array}$	$\begin{array}{r} 10 \\ \times 10 \\ \hline \end{array}$
$\begin{array}{r} 11 \\ \times 10 \\ \hline \end{array}$	$\begin{array}{r} 12 \\ \times 10 \\ \hline \end{array}$	$\begin{array}{r} 0 \\ \times 11 \\ \hline \end{array}$

$\begin{array}{r} 1 \\ \times 11 \\ \hline \end{array}$	$\begin{array}{r} 2 \\ \times 11 \\ \hline \end{array}$	$\begin{array}{r} 3 \\ \times 11 \\ \hline \end{array}$
$\begin{array}{r} 4 \\ \times 11 \\ \hline \end{array}$	$\begin{array}{r} 5 \\ \times 11 \\ \hline \end{array}$	$\begin{array}{r} 6 \\ \times 11 \\ \hline \end{array}$
$\begin{array}{r} 7 \\ \times 11 \\ \hline \end{array}$	$\begin{array}{r} 8 \\ \times 11 \\ \hline \end{array}$	$\begin{array}{r} 9 \\ \times 11 \\ \hline \end{array}$

10×11	11×11	12×11
0×12	1×12	2×12
3×12	4×12	5×12

$\begin{array}{r} 6 \\ \times 12 \\ \hline \end{array}$	$\begin{array}{r} 7 \\ \times 12 \\ \hline \end{array}$	$\begin{array}{r} 8 \\ \times 12 \\ \hline \end{array}$
$\begin{array}{r} 9 \\ \times 12 \\ \hline \end{array}$	$\begin{array}{r} 10 \\ \times 12 \\ \hline \end{array}$	$\begin{array}{r} 11 \\ \times 12 \\ \hline \end{array}$
$\begin{array}{r} 12 \\ \times 12 \\ \hline \end{array}$		

$1\overline{)2}$	$1\overline{)5}$	$1\overline{)8}$
$1\overline{)1}$	$1\overline{)4}$	$1\overline{)7}$
$1\overline{)0}$	$1\overline{)3}$	$1\overline{)6}$

$1\overline{)11}$	$2\overline{)2}$	$2\overline{)8}$
$1\overline{)10}$	$2\overline{)0}$	$2\overline{)6}$
$1\overline{)9}$	$1\overline{)12}$	$2\overline{)4}$

$2\overline{)14}$

$2\overline{)20}$

$3\overline{)0}$

$2\overline{)12}$

$2\overline{)18}$

$2\overline{)24}$

$2\overline{)10}$

$2\overline{)16}$

$2\overline{)22}$

$3\overline{)3}$	$3\overline{)6}$	$3\overline{)9}$
$3\overline{)12}$	$3\overline{)15}$	$3\overline{)18}$
$3\overline{)21}$	$3\overline{)24}$	$3\overline{)27}$

$3\overline{)36}$	$4\overline{)8}$	$4\overline{)20}$
$3\overline{)33}$	$4\overline{)4}$	$4\overline{)16}$
$3\overline{)30}$	$4\overline{)0}$	$4\overline{)12}$

$4\overline{)24}$	$4\overline{)28}$	$4\overline{)32}$
$4\overline{)36}$	$4\overline{)40}$	$4\overline{)44}$
$4\overline{)48}$	$5\overline{)0}$	$5\overline{)5}$

$5\overline{)20}$	$5\overline{)35}$	$5\overline{)50}$
$5\overline{)15}$	$5\overline{)30}$	$5\overline{)45}$
$5\overline{)10}$	$5\overline{)25}$	$5\overline{)40}$

$6\overline{)0}$	$6\overline{)18}$	$6\overline{)36}$
$5\overline{)60}$	$6\overline{)12}$	$6\overline{)30}$
$5\overline{)55}$	$6\overline{)6}$	$6\overline{)24}$

$6\overline{)54}$	$6\overline{)72}$	$7\overline{)14}$
$6\overline{)48}$	$6\overline{)66}$	$7\overline{)7}$
$6\overline{)42}$	$6\overline{)60}$	$7\overline{)0}$

$7\overline{)35}$

$7\overline{)56}$

$7\overline{)77}$

$7\overline{)28}$

$7\overline{)49}$

$7\overline{)70}$

$7\overline{)21}$

$7\overline{)42}$

$7\overline{)64}$

$8\overline{)8}$	$8\overline{)32}$	$8\overline{)56}$
$8\overline{)0}$	$8\overline{)24}$	$8\overline{)48}$
$7\overline{)84}$	$8\overline{)16}$	$8\overline{)40}$

$8\overline{)80}$

$9\overline{)0}$

$9\overline{)27}$

$8\overline{)72}$

$8\overline{)96}$

$9\overline{)18}$

$8\overline{)64}$

$8\overline{)88}$

$9\overline{)9}$

$9\overline{)54}$

$9\overline{)81}$

$9\overline{)108}$

$9\overline{)45}$

$9\overline{)72}$

$9\overline{)99}$

$9\overline{)36}$

$9\overline{)63}$

$9\overline{)90}$

$10\overline{)0}$	$10\overline{)10}$	$10\overline{)20}$
$10\overline{)30}$	$10\overline{)40}$	$10\overline{)50}$
$10\overline{)60}$	$10\overline{)70}$	$10\overline{)80}$

$10\overline{)110}$	$11\overline{)11}$	$11\overline{)44}$
$10\overline{)100}$	$11\overline{)0}$	$11\overline{)33}$
$10\overline{)90}$	$10\overline{)120}$	$11\overline{)22}$

11)77

11)110

12)0

11)66

11)99

11)132

11)55

11)88

11)121

$12\overline{)12}$

$12\overline{)24}$

$12\overline{)36}$

$12\overline{)48}$

$12\overline{)60}$

$12\overline{)72}$

$12\overline{)84}$

$12\overline{)96}$

$12\overline{)108}$

$12\overline{)144}$

$12\overline{)132}$

$12\overline{)120}$

VOCABULARY CARDS

Use the vocabulary cards to practice and review this year's new math terms. Suggestions for using the cards are in the Teacher's Edition, on the Chapter at a Glance page.

Consider having students organize their vocabulary cards in Math Words Files—containers made from zip-top bags or small boxes, such as crayon or computer disk boxes. Encourage students to consult their Math Word Files to confirm meanings, verify pronunciations, and check spellings.

To copy the cards, set the copy machine to 2-sided copies. Align the perforated edge with the left-hand (or top) guide on the glass and copy. Flip the page and align the perforated edge with the opposite (right-hand or bottom) guide. Copy.

Pronunciation Key

a	**a**dd, m**a**p	h	**h**ope, **h**ate	ô	**o**rder, j**aw**	t͟h	**th**is, ba**th**e
ā	**a**ce, r**a**te	i	**i**t, give	oi	**oi**l, b**oy**	u	**u**p, d**o**ne
â(r)	c**a**re, **a**ir	ī	**i**ce, wr**i**te	ou	p**ou**t, n**ow**	û(r)	b**ur**n, t**er**m
ä	p**a**lm, f**a**ther	j	**j**oy, le**dg**e	o͝o	t**oo**k, f**u**ll	yo͞o	f**u**se, f**ew**
b	**b**at, ru**b**	k	**c**ool, ta**k**e	o͞o	p**oo**l, f**oo**d	v	**v**ain, e**v**e
ch	**ch**eck, cat**ch**	l	**l**ook, ru**l**e	p	**p**it, sto**p**	w	**w**in, a**w**ay
d	**d**og, ro**d**	m	**m**ove, see**m**	r	**r**un, poo**r**	y	**y**et, **y**earn
e	**e**nd, p**e**t	n	**n**ice, ti**n**	s	**s**ee, pa**ss**	z	**z**est, mu**s**e
ē	**e**qual, tr**ee**	ng	ri**ng**, so**ng**	sh	**sh**ure, ru**sh**	zh	vi**s**ion, plea**s**ure
f	**f**it, hal**f**	o	**o**dd, h**o**t	t	**t**alk, si**t**		
g	**g**o, lo**g**	ō	**o**pen, s**o**	th	**th**in, bo**th**		

ə the schwa, an unstressed vowel representing the sound spelled *a* in **a**bove, *e* in sick**e**n, *i* in poss**i**ble, *o* in mel**o**n, *u* in circ**u**s

Other symbols:
- • separates words into syllables
- ′ indicates stress on a syllable

benchmark	**period**
millions	**expression**
variable	**equation**
survey	**cumulative frequency**

pir′ē•əd **Each group of three digits in a number (6)**	bench′märk **A known number of things that helps you understand the size or amount of a different number of things (2)**
ik•spre′shən **A part of a number sentence that has numbers and operation signs but does not have an equal sign (54)**	mil′yənz **The period after thousands, equal to one thousand thousands (8)**
i•kwā′zhən **A number sentence which states that two amounts are equal (61)**	vâr′ē•ə•bəl **A letter or symbol that stands for any number (60)**
kyōō′myə•lə•tiv frē′kwən•sē **The sum of the frequency of data as they are collected; a running total of items being counted (83)**	sûr′vā **A method of gathering information by asking questions and recording people's answers (82)**

frequency	mode
median	line plot
range	outlier
leaf	stem-and-leaf plot

mōd **The number or item that occurs most often in a set of data (86)**	frē′kwen•sē **The number of times a response occurs (83)**
līn plät **A graph that shows the frequency of data along a number line (88)**	mē′dē•ən **The middle number in an ordered set of data (86)**
out′lī•ər **A value separated from the rest of the data (88)**	rānj **The difference between the greatest and the least values in a set of data (88)**
stem ənd lēf plät **A table that shows groups of data arranged by place value (90)**	lēf **A ones digit in a stem-and-leaf plot (90)**

stem	**scale**
interval	**double-bar graph**
line graph	**trends**
second (sec)	**A.M.**

skāl **A series of numbers placed at fixed distances on a graph to help label the graph (92)**	stem **A tens digit in a stem-and-leaf plot (90)**
du′bəl bär graf **A graph used to compare similar kinds of data (100)**	in′tər•vəl **The distance between two numbers on the scale of a graph (92)**
trendz **On a graph, areas where the data increase, decrease, or stay the same over time (105)**	līn graf **A graph that uses a line to show how data change over a period of time (102)**
ā•em **Between midnight and noon (118)**	se′kənd **A small unit of time; 60 seconds = 1 minute (116)**

P.M.	elapsed time
century	decade
fact family	inverse operation
Grouping Property of Multiplication	multistep problem

i•lapst′ tīm **The time that passes from the start of an activity to the end of that activity (120)**	pē•əm **After noon (118)**
de′kād **A measure of time, equal to 10 years (127)**	sen′chə•rē **A measure of time on a calendar, equal to 100 years (127)**
in′vərs ä•pə•rā′shənz **Operations that undo each other, like addition and subtraction, or multiplication and division (140)**	fakt fam′ə•lē **A set of related multiplication and division, or addition and subtraction, equations using the same numbers (140)**
mul′ti•stəp prä′bləm **A problem requiring more than one step to solve (234)**	grōō•ping prä′pər•tē əv mul•tə•plə•kā′shən **The property stating that you can group factors in different ways and still get the same product (150)**

remainder	**compatible numbers**
mean	**multiple**
composite number	**prime number**
prime factor	**factor tree**

kəm•pa′tə•bəl num′bərz

Nearby numbers that are easy to compute mentally (258)

ri•mān′dər

The amount left over after you find a quotient (246)

mul′tə•pəl

The product of a given whole number and another whole number (298)

mēn

The average of a set of numbers, found by dividing the sum of the set by the number of addends (276)

prīm num′bər

A number that has only two factors: 1 and itself (302)

kəm•pä′zət num′bər

A number that has more than two factors (302)

fak′tər trē

A diagram that shows the prime factors of a number (306)

prīm fak′tər

A factor that is a prime number (306)

line	line segment
plane	point
ray	right angle
acute angle	obtuse angle

līn seg′mənt **A part of a line that has two endpoints (320)**	līn **A straight path in a plane, extending in both directions with no endpoints (320)**
point **A location on an object or in space (320)**	plān **A flat surface that extends without end in all directions (320)**
rīt an′gəl **An angle that forms a square corner and has a measure of 90° (321)**	rā **A part of a line; begins at one endpoint and goes on forever in one direction (320)**
əb•to͞os′ an′gəl **An angle that has a measure greater than a right angle (greater than 90°) (321)**	ə•kyo͞ot′ an′gəl **An angle that has a measure less than a right angle (less than 90°) (321)**

vertex	parallel lines
perpendicular lines	intersecting lines
congruent	transformation
slide	flip

pâr′ə•lel līnz **Lines that never intersect (324)**	vûr′teks **The point at which two rays of an angle or two or more line segments meet in a plane figure, or where three or more sides meet in a solid figure (321)**
in•tər•sek′ting līnz **Lines that cross each other at exactly one point (324)**	pər•pən•di′kyə•lər līnz **Lines that intersect to form four right angles (324)**
trans•fər•mā′shən **The movement of a figure by a slide, flip, or turn (326)**	kən•grōō′ənt **Having the same size and shape (326)**
flip **A movement of a figure to a new position by flipping the figure over a line (326)**	slīd **A movement of a figure to a new position without turning or flipping it (326)**

turn	similar
line symmetry	rotational symmetry
degree	protractor
radius	center

si′mə•lər **Having the same shape as something else but possibly different in size (326)**	tûrn **A movement of a figure to a new position by rotating the figure around a point (326)**
rō•tā′shən•əl si′mə•trē **What a figure has if it can be turned about a central point and still look the same in at least two positions (330)**	līn si′mə•trē **What a figure has if it can be folded about a line so that its two parts match exactly (330)**
prō′trak•tər **A tool for measuring the size of an angle (340)**	di•grē′ **The unit used for measuring angles or temperatures (338)**
sen′tər **A point inside a circle from which all points on the circle are the same distance (342)**	rā′dē•əs **A line segment with one endpoint at the center of a circle and the other endpoint on the circle (342)**

chord	circle
compass	diameter
circumference	isosceles triangle
scalene triangle	equilateral triangle

sər′kəl **A closed figure made up of points that are the same distance from the center point (342)**	kôrd **A line segment with endpoints on a circle (342)**
dī•a′mə•tər **A chord passing through the center of a circle (342)**	kəm′pəs **A tool used to construct circles (342)**
ī•sä′sə•lēz′ trī′an•gəl **A triangle with only two congruent sides (346)**	sər•kum′fər•əns **The distance around a circle (344)**
ē•kwə•la′tə•rəl trī′an•gəl **A triangle with three congruent sides (346)**	skā′lēn trī′an•gəl **A type of triangle with no congruent sides (346)**

parallelogram	rhombus
trapezoid	Venn diagram
fraction	equivalent fractions
simplest form	mixed number

räm′bəs **A parallelogram with four congruent sides and with opposite angles that are congruent (349)**	pâr•ə•lel′ə•gram **A quadrilateral whose opposite sides are parallel and congruent (349)**
ven dī′ə•gram **A diagram that shows relationships among sets of things (352)**	tra′pə•zoid **A quadrilateral with only one pair of parallel sides (349)**
ē•kwiv′ə•lənt frak′shənz **Two or more fractions that name the same amount (366)**	frak′shən **A number that names a part of a whole (364)**
mikst num′bər **An amount given as a whole number and a fraction (378)**	sim′pləst fôrm **A fraction is in simplest form when 1 is the only number that can divide evenly into the numerator and the denominator (368)**

like fractions	**unlike fractions**
decimal	**thousandth**
equivalent decimals	**linear units**
inch (in.)	**foot (ft)**

un′līk frak′shənz

Fractions with different denominators (398)

līk frak′shənz

Fractions with the same denominator (386)

thou′zəndth

One of one thousand equal parts (410)

de′sə•məl

A number with one or more digits to the right of the decimal point (406)

li′nē•ər yo͞o′nəts

Units which measure length, width, height, or distance (452)

ē•kwiv′ə•lənt de′sə•məlz

Two or more decimals that name the same amount (412)

fo͝ot

A customary unit used for measuring length; 1 foot = 12 inches (452)

inch

A customary unit used for measuring length (452)

yard (yd)	**mile (mi)**
capacity	**cup (c)**
pint (pt)	**gallon (gal)**
quart (qt)	**weight**

mīl **A customary unit for measuring length; 5,280 feet = 1 mile (452)**	yärd **A customary unit for measuring length; 3 feet = 1 yard (452)**
kup **A customary unit for measuring capacity; 8 ounces = 1 cup (460)**	kə•pa′sə•tē **The amount a container can hold when it is filled (460)**
ga′lən **A customary unit for measuring capacity; 4 quarts = 1 gallon (460)**	pīnt **A customary unit for measuring capacity; 2 cups = 1 pint (460)**
wāt **How heavy an object is (462)**	kwôrt **A customary unit for measuring capacity; 2 pints = 1 quart (460)**

ounce (oz)	**pound (lb)**
ton (T)	**kilometer (km)**
meter (m)	**centimeter (cm)**
decimeter (dm)	**milliliter (mL)**

pound **A customary unit used for measuring weight; 16 ounces = 1 pound (462)**	ouns **A customary unit used for measuring weight; 16 ounces = 1 pound (462)**
kə•lä′mə•tər **A unit of length in the metric system; 1,000 meters = 1 kilometer (470)**	tun **A customary unit used for measuring weight; 2,000 pounds = 1 ton (462)**
sən′tə•mē•tər **A unit of length in the metric system; 100 centimeters = 1 meter (470)**	mē′tər **A unit of length in the metric system; 100 centimeters = 1 meter (470)**
mi′lə•lē•tər **A unit of capacity in the metric system; 1,000 milliliters = 1 liter (476)**	de′sə•mē•tər **A unit of length in the metric system; 10 decimeters = 1 meter (470)**

liter (L)	kilogram (kg)
mass	gram (g)
pentagon	quadrilateral
triangle	perimeter

ki′lə•gram **A unit of mass in the metric system; 1 kilogram = 1,000 grams (478)**	lē′tər **A unit of capacity in the metric system; 1 liter = 1,000 milliliters (476)**
gram **A unit of mass in the metric system; 1,000 grams = 1 kilogram (478)**	mas **The amount of matter in an object (478)**
kwä•drə•la′tə•rəl **A polygon with four sides and four angles (486)**	pen′tə•gän **A polygon with five sides and five angles (486)**
pə•ri′mə•tər **The distance around a figure (486)**	trī′an•gəl **A polygon with three sides and three angles (486)**

formula	**hexagon**
octagon	**polygon**
regular polygon	**area**
two-dimensional	**three-dimensional**

hek′sə•gän **A polygon with six sides and six angles (486)**	fôr′myə•lə **A set of symbols that expresses a mathematical rule (486)**
pä′lē•gän **A closed plane figure with straight sides; Each side is a line segment. (486)**	äk′tə•gän **A polygon with eight sides and eight angles (486)**
âr′ē•ə **The number of square units needed to cover a surface (492)**	reg′yə•lər pä′lē•gän **A polygon that has sides that are the same length (486)**
thrē•də•mən′shən•əl **Measured in three directions, such as length, width, and height (508)**	to͞o•də•mən′shən•əl **Measured in two directions, such as length and width (508)**

net	volume
cubic unit	outcomes
event	tree diagram
predict	likely

väl′yəm **The measure of the space a solid figure occupies (514)**	net **A two-dimensional pattern of a three-dimensional figure (512)**
out•kəmz **The possible results of an experiment (528)**	kyōō′bik yōō′nət **A unit of volume with dimensions of 1 unit × 1 unit × 1 unit (514)**
trē̄ dī′ə•gram **An organized list that shows all possible outcomes for an event (530)**	i•vent′ **One outcome or a combination of outcomes in an experiment (528)**
līk′lē **Having a greater than even chance of happening (534)**	pri•dikt′ **To tell what might happen (534)**

unlikely	equally likely
mathematical probability	fairness
degrees Fahrenheit (°F)	degrees Celsius (°C)
opposites	negative numbers

ē′kwə•lē lī′klē **Having the same chance of happening as something else (534)**	un•lī′klē **Having a less than even chance of happening (534)**
fâr′nəs **Fairness in a game means that one player is as likely to win as another. Each player has an equal chance of winning. (546)**	math•ma′ti•kəl prä•bə•bi′lə•tē **A comparison of the number of favorable outcomes to the number of possible outcomes of an event (542)**
di•grēz səl′sē•us **A standard unit for measuring temperature in the metric system (556)**	di•grēzs fâr′ən•hīt **A standard unit for measuring temperature in the customary system (554)**
ne′gə•tiv num′bərz **All the numbers to the left of zero on the number line (558)**	ä′pə•zət **Numbers on the number line which are the same distance from 0 (558)**

y-coordinate	**y-axis**
ordered pair	**x-coordinate**
x-axis	**function table**

wī′ak′səs

The vertical line on a coordinate grid (568)

wī′kō•ôrd′ə•nət

The second number in an ordered pair; It tells the distance to move up or down. (568)

eks′kō•ôrd′ə•nət

The first number in an ordered pair; It tells the distance to move horizontally. (568)

ôr′dərd pâr

A pair of numbers used to locate a point on a coordinate grid. The first number tells how far to move horizontally, and the second number tells how far to move vertically. (568)

funk′shən tā′bəl

A table that matches each input value with an output value. The output values are determined by the function (570)

eks′ak′səs

The horizontal line on a coordinate grid (568)